Green Energy and Technology

More information about this series at http://www.springer.com/series/8059

María Isabel Roldán Serrano

Concentrating Solar Thermal Technologies

Analysis and Optimisation by CFD Modelling

Springer

María Isabel Roldán Serrano
CIEMAT-Plataforma Solar de Almería
Tabernas
Spain

ISSN 1865-3529 ISSN 1865-3537 (electronic)
Green Energy and Technology
ISBN 978-3-319-45882-3 ISBN 978-3-319-45883-0 (eBook)
DOI 10.1007/978-3-319-45883-0

Library of Congress Control Number: 2016955527

Printed on acid-free paper

This Springer imprint is published by Springer Nature
The registered company is Springer International Publishing AG
The registered company address is: Gewerbestrasse 11, 6330 Cham, Switzerland

Acknowledgements

This book is partially supported by CIEMAT-Plataforma Solar de Almería which is the largest European center for research, development, and testing of concentrating solar technologies. Special thanks go to the Solar Concentrating Systems Unit which is directly involved in the research and development of this technology.

The author also gratefully acknowledges the information supplied by the International Energy and Environment Foundation, as well as the helpful comments and suggestions of the reviewers, which have improved this work.

Contents

1 Introduction 1
- 1.1 Global Energy Situation 1
- 1.2 Energy Consumption Trends 2
  - 1.2.1 China 3
  - 1.2.2 India 3
  - 1.2.3 United States 3
  - 1.2.4 European Union 4
  - 1.2.5 Russia 4
  - 1.2.6 Mexico 5
  - 1.2.7 Japan 5
  - 1.2.8 Southeast Asia 6
  - 1.2.9 Middle East 6
  - 1.2.10 Latin America 7
  - 1.2.11 Africa 7
- 1.3 Renewable Energy Situation 8
- 1.4 Solar Thermal Electricity 9
- References 10

2 Concentrating Solar Thermal Technologies 11
- 2.1 Introduction 11
- 2.2 Concentration of Solar Radiation 12
- 2.3 Concentrating Solar Thermal (CST) Technologies 14
  - 2.3.1 Parabolic Trough Collectors 16
  - 2.3.2 Linear Fresnel 17
  - 2.3.3 Solar Tower 18
  - 2.3.4 Parabolic Dish 20
  - 2.3.5 Solar Furnace 21
- 2.4 Status of CST Technologies 22
  - 2.4.1 Medium-Concentration Solar Technology 22

2.4.2 High-Concentration Solar Technology. 22
2.4.3 Performance of CST Technologies 23
References. 24

3 Computational Fluid Dynamics in Green Design 25
3.1 Introduction 25
3.2 Green Engineering 26
3.3 Computational Fluid Dynamics (CFD) 27
3.3.1 Advantages of CFD Simulation. 28
3.3.2 Disadvantages of CFD Simulation 29
3.4 Principles of CFD Simulation. 30
3.4.1 Pre-processing. 30
3.4.2 Solver 31
3.4.3 Post-processing 33
3.5 Governing Equations for Fluid Flow 33
3.6 Solving by CFD 34
3.7 CFD Application in Green Design 36
3.7.1 Green Building Design 36
3.7.2 Wind Power 37
3.7.3 Concentrating Solar Thermal Technologies. 38
References. 38

4 Optimisation of Solar Receivers. 39
4.1 Introduction 39
4.2 Solar Receivers. 40
4.2.1 Solar Receivers in Medium-Concentration Technologies. 40
4.2.2 Solar Receivers in High-Concentration Technologies 40
4.3 Optimisation of Solar Receivers 43
4.3.1 Thermal Analysis and Optimisation of the Receiver Design. 43
4.3.2 Operating Conditions 43
4.4 Thermal Analysis and Optimisation of a Solar Volumetric Receiver by CFD Simulation 44
4.4.1 Thermal Analysis of a Solar Volumetric Receiver 44
4.4.2 Design Optimisation for a Solar Volumetric Receiver. 49
4.5 Optimisation of the Operating Conditions for a Solar Volumetric Receiver by CFD Simulation 57
4.5.1 Facility Description. 57
4.5.2 Procedure 58
4.5.3 Numerical Modelling 60
4.5.4 Geometry Definition and Mesh Design. 61
4.5.5 Boundary Conditions 63

4.5.6 Results and Discussion 65
4.5.7 Summary and Conclusions 70
References. 71

5 Heat Transfer Fluids Used in Concentrating Solar Thermal Technologies 73
5.1 Introduction 73
5.2 Heat Transfer Fluids for CST Systems 74
5.3 Heat Transfer Fluids Used in Commercial CST Plants 75
5.3.1 Thermal Oils 75
5.3.2 Organics 75
5.3.3 Molten Salts 76
5.3.4 Water/Steam 76
5.4 Alternative Heat Transfer Fluids in CST Plants 77
5.4.1 Air and Other Gases 77
5.4.2 Liquid Metals 77
5.4.3 Solid Particles 78
5.4.4 Supercritical Fluids 78
5.5 CFD Analysis of a Tubular Receiver Design for Two Different HTFs 79
5.5.1 Receiver Description 79
5.5.2 Procedure 79
5.5.3 CFD Modelling 81
5.5.4 Results and Discussion 82
5.5.5 Summary and Conclusions 84
References. 85

6 Conclusions 87
6.1 Summary and Conclusions 87

Chapter 1
Introduction

In order to address the global challenges of clean energy, climate change and sustainable development, it is necessary to accelerate the development of environmentally-friendly energy technologies. In this context, concentrating solar thermal technologies are increasingly relevant because of the need to reduce carbon dioxide emissions in electricity and heat generation required to reach the goal of limiting climate change to 2 °C above pre-industrial levels. Therefore, this chapter aims to remark the importance of developing concentrating solar thermal technologies adapted to the requirements of the current energy supply.

1.1 Global Energy Situation

Any modern society to develop requires an energy system that ensures a constant supply. This system should be based on abundant resources of an adequate quality, easily portable and obtained at a cost-efficient price. Throughout the history of humanity, two main criteria have been followed to choose energy systems: technical availability and economic viability. However, during the last few decades, the environmental impact of the energy system has become another important aspect to consider. This new factor is a key point in assessing and comparing different energy sources and technologies for a country's energy grid (Cartelle et al. 2015).

Energy production and use contribute to around two-thirds of global greenhouse gas (GHG) emissions, mainly carbon dioxide (CO_2) emissions (see Table 1.1), which imply that actions in the energy sector are crucial to addressing the climate change challenge. Therefore, a transformation of the world's energy system to a safe and sustainable low-carbon one is underway. In fact, renewable energies accounted for nearly half of the growth in global electricity generation capacity in 2014. Nowadays, energy efficiency regulations cover 27 % of the world's energy consumption, up from 12 % in 2005. Furthermore, the rise of distributed generation, smart grids and storage technologies are rapidly changing the way energy is supplied

M.I. Roldán Serrano, *Concentrating Solar Thermal Technologies*,
Green Energy and Technology, DOI 10.1007/978-3-319-45883-0_1

Table 1.1 Global energy- and process-related greenhouse-gas emissions in the intended nationally determined contributions (INDCs) Scenario

Year	2013	2020	2025	2030
Energy-related ($Gt\ CO_2{-}eq$)				
Carbon dioxide (CO_2)	32.2	33.9	34.4	34.8
Methane (CH_4)	3.0	3.1	3.1	3.1
Nitrous oxide (N_2O)	0.3	0.3	0.4	0.4
Process-related ($Gt\ CO_2{-}eq$)				
Carbon dioxide (CO_2)	2.0	2.2	2.2	2.3
Total	37.5	39.5	40.0	40.6

Source International Energy Agency (IEA-b 2015)

and consumed. Thus, safe and sustainable low-carbon technologies and energy efficiency play an important role in reducing GHG emissions to hold the increase in global average temperature below 2 °C above pre-industrial levels (IEA-a 2015).

Five key actions have been identified to reduce GHG emissions while maintaining economic growth (IEA-b 2015):

- Increasing energy efficiency in the industry, buildings and transport sectors
- Reducing the use of the least-efficient coal-fired power plants and banning their construction
- Increasing investment in renewable energy technologies in the power sector
- Gradual phasing out of fossil-fuel subsidies to end-users by 2030
- Reducing methane emissions in oil and gas production.

The adoption of these measures can lock-in the recently observed decoupling emissions growth from economic one. In fact, the energy sector can achieve a peak in GHG emissions by around 2020, 39.5 gigatonnes of carbon-dioxide equivalent ($Gt\ CO_2{-}eq$) according to Table 1.1, while maintaining the same level of economic growth and development. Thus, a near-term peak in global emissions will drive governments to transform their energy economies. Moreover, a clear and credible vision of long-term decarbonisation is crucial to provide the right signals for investment and to allow a low-carbon, high-efficiency energy sector to be at the core of international efforts to combat climate change.

1.2 Energy Consumption Trends

Energy use worldwide is set to grow by one-third to 2040, which is driven primarily by India, China, Africa, the Middle East and Southeast Asia. However, declines are led by the European Union (−15 % over the period to 2040), Japan (−12 %) and the United States (−3 %). Furthermore, electricity gains importance in many end-use sectors, making up almost a quarter of final energy consumption by 2040. Therefore, the power sector tends to lead the change towards a decarbonised energy system (IEA-c 2015).

1.2.1 China

China is the world's largest producer and consumer of coal and deploys more renewable power generation capacity than any other country. Furthermore, by the 2030s, it will become the biggest consumer of oil (more than the United States) and will have a larger gas market than the European Union. China's total energy demand in 2040 is expected to be almost double that the United States' one. Therefore, its government has proposed policies of energy efficiency, and less economic and population growth. In fact, half of China's energy use today is already subject to mandatory efficiency standards and to continued improvements in efficiency, alongside large-scale deployment of wind, solar, hydro and nuclear power. Hence, it is expected that China decouples its economic expansion from emissions growth by around 2020 (IEA-c 2015).

1.2.2 India

India is the third-largest economy, but accounts for only 6 % of global energy use and around 240 million people still lack access to electricity. Therefore, new policies are conducted to accelerate the country's modernisation and develop its manufacturing base. Thus, India is entering a sustained period of rapid growth in energy consumption, becoming the largest source of growth in global coal use and increasing the oil demand by more than in any other country. A rapidly expanding energy sector could intensify serious challenges with water stress and local air pollution, but these risks can be avoided by integrated policies on land use and urbanisation, pollution controls, technology development, and a restless focus on energy efficiency. In order to achieve a 40 % share of non-fossil fuel capacity in the power sector by 2030, India also steps up its deployment of low-carbon technologies focusing on solar and wind power (IEA-c 2015).

1.2.3 United States

The collective intention to reach a low-carbon global economy as rapidly as possible has kept the United States on a path to achieving 80 % reductions or more by 2050. Energy and climate policies introduced in recent years are already having a material impact on the projected emissions trajectory for the United States. Energy-related CO_2 emissions accounted for all of the net reductions, with around 60 % coming from the power sector, mainly as a result of lower natural gas prices encouraging coal-to-gas switching, increasing contributions from renewable energies and, to a much lesser extent, coal plants being retired in anticipation of environmental regulations.

Furthermore, electricity demand in the United States increases by just 9 % from 2013 to 2025, as energy efficiency efforts slow growth over time. The decline of CO_2 emissions expected in power sector by 2025 is largely driven by climate policies and efforts to improve both power sector efficiency and end-use efficiency (reducing electricity demand). Coal-fired capacity decreases by more than 20 % by 2025, as new additions are limited to the highest efficiency, to carbon capture and storage-ready technologies, and to the replacement of the oldest and least-efficient existing capacity (IEA-b 2015).

1.2.4 European Union

The European Union (EU)'s 2030 framework for energy and climate policies sets out targets to increase the share of renewable energy to at least 27 % of final energy consumption and to improve energy efficiency by at least 27 % relative to a projected reference level in 2030. As a consequence of the 2030 framework, CO_2 emissions from the energy sector will fall by around 37 % and non-CO_2 greenhouse gases by around 55 %. This framework builds on the target to reduce EU GHG emissions by 20 % by 2020, closer to the EU's long-term objective of cutting emissions by at least 80 % by 2050.

This implies the reorientation of the EU's energy system with the share of low-carbon energy sources growing from 27 % of primary energy demand in 2013 to 37 % in 2030. As the share of fossil fuels declines, the relative weighting between the fossil fuels also moves towards gas. Natural gas imports grow by around 18 % from 2013 to 2030 and the role of nuclear power declines in some countries, pulling its share of the regional energy mix down slightly over time. While the EU's potential for large hydropower has already been largely harnessed, there is a major expansion in its use of wind, solar and bioenergy.

Electricity demand in the European Union increases by around 10 % in 2030 mainly due to the power consumption of the service sector, and industrial demand starts to decline in the 2020s. Renewables account for more than half of the European Union's power generation capacity in 2030. Coal-fired capacity declines by nearly 40 % and the carbon intensity of the EU's power sector halves by 2030. However, the gas-fired capacity increases by one-third (IEA-b 2015).

1.2.5 Russia

Russia's energy-related CO_2 emissions decrease slightly from 2013 to 2030 and the carbon intensity of Russia's economy declines through to 2030. However, it remains high in comparison with many other countries (Fig. 1.1) due to the continuing high share of fossil fuels in the energy mix (led by natural gas), the inefficient use of energy in some sectors and because of the climatic conditions. The efficiency of

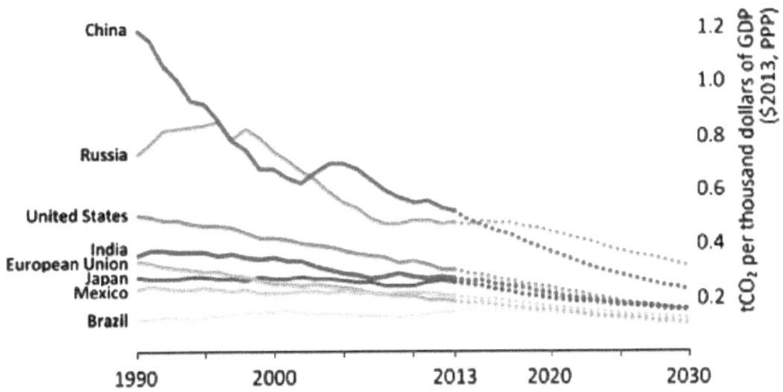

Fig. 1.1 Carbon intensity of the economy by selected region in the INDCs scenario (IEA-b 2015)

Russia's thermal electricity generation is projected to increase significantly to 2030 together with the share of nuclear in the electricity mix. These factors help to suppress growth in power sector emissions. Russia is one of the world's major oil and gas producers and its methane emissions mainly come from these sectors, although these emissions tend to decline between 2013 and 2030 (IEA-b 2015).

1.2.6 Mexico

Mexico has already established a general strategy based on the climate change mitigation and the assessment of its own vulnerability to climate change. This country aims to achieve an unconditional 25 % reduction in 2030 considering its emissions of greenhouse gases and short-lived climate pollutants. The energy sector accounts for the majority of GHG emissions, with the next largest source being methane emissions. In 2030, Mexico's energy-related CO_2 emissions tend to increase, while its economy is expected to be almost double in size and its population growth by around one-fifth over the same time period. Thus, electricity demand increases by around 50 % by 2030, mainly as a result of higher industrial activity. Despite the significant rise in electricity demand, CO_2 emissions from power generation decrease slightly to 2030, as natural gas- and renewable-based generation grows and oil-fired generation is drastically reduced. The share of electricity generation from low-carbon sources increases to nearly 40 %, with wind and hydropower as the largest sources, followed by geothermal, nuclear and solar PV ones (IEA-b 2015).

1.2.7 Japan

The energy strategy proposed by Japan considers nuclear power as an important source of base load electricity, but the process of regulatory approval is still far from complete and it has boosted Japan's consumption of fossil fuels and stepped up its reliance on energy imports. However, it is expected that Japan's energy-related CO_2 emissions are around 25 % lower than 2013 levels by 2030 in the INDC Scenario, increasing its economy by 17 % over the same period. Most idled nuclear power plants steadily return to service, after receiving regulatory approval, and in parallel renewable energies (led by solar one and, to a lesser extent, wind) continue to grow in the power mix.

In this context, the use of fossil fuels in the power sector declines from around 85 % of generation today to about 55 % by 2030. In transport, energy efficiency efforts, together with policies to support alternative fuels, help push oil demand down by one-third by 2030. Actions that contribute to reduce Japan's fossil-fuel demand bring benefits not only in terms of emissions, but also in terms of energy security, reducing Japan's oil-import bills by more than one-fifth in 2030 with respect to 2013 (IEA-b 2015).

1.2.8 Southeast Asia

Energy-related CO_2 emissions in Southeast Asia increase by 60 % from today's level to the one reached in 2030, with Indonesia accounting for around 35 % of the total. The emission growth reflects the greater share of fossil fuels in the energy mix, and the largest contribution comes from the power sector as a result of the rapid rise in demand for electricity and coal. The growing importance of coal is mainly due to Indonesia's abundant and low-cost coal supply, contrasting with relatively expensive natural gas in most countries. Indonesia, Thailand and Malaysia are the largest economies in this area whose industrialisation pushes industrial emissions up by more than 40 % by 2030 (IEA-b 2015).

1.2.9 Middle East

Energy demand has grown strongly in the Middle East in recent years, but the most significant contribution is due to the major energy exporters (such as Saudi Arabia, Qatar, Kuwait and others) in the region. Thus, energy-related CO_2 emissions rise by around 35 % in 2030, and per-capita emissions are projected to reach around double the world average at that time. As a consequence, there is a spiralling energy demand growth that it is difficult to control by policy-makers. Emission increase is due not only to low energy prices that encourage the growth of energy-intensive industries (foremost the petrochemical industry), but also to wasteful use of energy because of the extent of fossil-fuel subsidies. In fact, average fuel consumption per vehicle in the Middle East is projected to remain the highest in the world in 2025.

Despite electricity demand increasing by around 75 % from 2013 to 2030, emissions grow by only around 15 %, as the power sector shifts from inefficient oil-fired power plants to gas-fired power plants and low-carbon technologies. Some countries in the region have targets and policies in place to expand low carbon sources in the power sector (such as Saudi Arabia, Kuwait, and Dubai), but progress has typically been limited so far. It is expected a significant growth in both oil and gas production in 2030. Efforts, such as those in southern Iraq, are underway to try, capture and utilise associated gas in power generation, petrochemicals and industry (IEA-b 2015).

1.2.10 Latin America

In Latin America, GHG emissions have historically been dominated by land use, land-use change and forestry, and agriculture. Energy-related CO_2 emissions in this region are significantly lower than the global average due not only to the dominance of low-carbon technologies in the electricity mix (particularly hydropower in Brazil), but also to the high share of biofuels in the transport sector. However, energy sector is becoming a more significant source of emission growth and it will be important in determining whether Latin America can maintain its low-carbon profile as domestic energy demand increases rapidly. Thus, the emission growth is largely driven by rising industrial activity.

In contrast, emissions from electricity generation decrease by around 7 % to 2030, as the power sector is able to meet additional demand for electricity through increased generation from hydropower, natural gas and wind. Since 2005, Brazil (the region's largest emitter) tries to implement policies to slow down deforestation, with a particular focus on how to contain growth in energy-related emissions. In 2008, this country announced a strategy to increase the efficiency of road vehicles; but these measures are not yet sufficient to tap into the large unrealised energy efficiency potential (IEA-b 2015).

1.2.11 Africa

Africa currently accounts for a small share of global GHG emissions, with a South Africa's contribution of more than one-third of its total. This region could suffer severe impacts from a changing climate, being Gabon the first African nation that has committed to keeping 2025 emissions at least 50 % below a business-as-usual level. Its strategy focuses primarily on land management, but also on reducing flaring from the oil and gas sector, boosting energy efficiency, increasing hydropower supply and setting up a domestic carbon offset market.

Electricity consumption in Africa doubles from 2013 to 2030. Around 500 million people are projected to gain access to electricity for the first time by 2030, and the regions' small industrial base expands significantly. Despite this, access to reliable, affordable modern energy remains a major challenge in many parts of the continent, repressing economic and social development.

African energy-related CO_2 emissions are projected to increase by around 40 % to 2030. Emissions in South Africa are expected to follow a peak in the period from 2020 to 2025, a plateau for a several years and then emissions start to decline in the 2030s. Key drivers of this trend are improved energy efficiency in end-use sectors and less dependence on coal for the power sector, as it turns towards renewables and nuclear (IEA-b 2015).

1.3 Renewable Energy Situation

The key actions proposed to reduce GHG emissions (Sect. 1.1) have important implications for the global energy mix, putting a brake on growth in oil and coal use within the next five years and further boosting renewables. In fact, despite the lower fossil-fuels prices of the last two years, the global investment in renewable-based power generation has continued and positive policy moves have been promoted in many countries. For instance, India has focused on the installation of non-hydro renewable energy capacity of 175 gigawatts (GW) by 2022 (of which solar PV is 100 GW).

Renewables-based power generation capacity is estimated to have increased by 128 GW in 2014, of which 37 % is wind power, almost one-third solar power and more than a quarter from hydropower. This became more than 45 % of world power generation capacity additions in 2014, consistent with the general increasing trend in recent years (IEA-b 2015).

The growth in wind capacity continued to be led by onshore installations, although offshore ones have also grown rapidly. China remains the largest wind power market, with 20 GW of new capacity, while Germany installed more than 5 GW of wind capacity in 2014. Relatively high (but declining) costs for offshore wind and delays in the build-up of grid connections have resulted in postponements of projects in some countries or in the cutting of capacity targets. However, wind leads the growth in renewables-based capacity in the United States, averaging over 6 GW per year to 2025, while solar technologies average around 5 GW per year, with solar photovoltaic (PV) accounting for the vast majority.

In fact, solar photovoltaic (PV) expanded strongly in Asia, particularly in China and Japan that has plans to increase to 20 % the proportion of electricity generated by renewables in 2030. Japan's policy actions are based on attractive feed-in tariffs, which have, in recent years, stimulated a supply of new solar-based electricity, but this has prompted concerns on the part of the utilities regarding the reliability of this supply and the stability of the grid.

Lower oil prices proved to be a challenge for other forms of renewable energy, including biofuels in transport and renewable heat, as the later competes directly with natural gas heating with a price normally linked to the oil one. While biofuels face challenges arising from lower oil prices, some other developments served to improve their outlook. In order to improve prospects for biofuels in Brazil, the government increased the ethanol blending rate from 25 to 27 % and that for biodiesel from 5 to 7 % together with the increase of gasoline taxes, while Argentina and Indonesia raised their biofuel mandates.

Africa presents a new trend for renewable energies. It starts to unlock its vast renewable energy resources, with half of the growth in power generation capacity coming from renewables. New hydropower capacity plays a major role in bringing down the region's average cost of power supply. Other renewables, led by solar technologies, make a growing contribution to supply, while geothermal is an important source of power in East Africa.

In European Union, it is expected that around one-quarter of total power generation by 2025 comes from variable renewables (wind and solar), highlighting the need to invest in greater levels of interconnection to handle the variability of supply, as well as in upgraded distribution networks. The effective technical and market integration of variable renewables with other forms of supply will be an important future challenge for the EU, as for many countries around the world. The EU strategy proposes increased interconnection across national markets, as a means to support such integration and energy security more generally.

1.4 Solar Thermal Electricity

As previously mentioned, it is projected that CO_2 emissions from energy sector increase by 61 % over 2011 levels by 2050. A range of technology solutions could contribute to preventing this increase, such as the improvement of the energy efficiency, renewable energies, nuclear power and the near-decarbonisation of fossil fuel-based power generation. However, the increasing investment in renewable-based power generation envisages slower deployment of nuclear and carbon capture and storage technologies, and more rapid expansion of renewables, notably solar and wind energy (IEA-d 2014).

Solar and wind energy have been pioneering renewable energy technologies, after hydropower. It is expected that the increased energy demand will come from the countries with high solar resource. The advantage is that solar energy can be exploited by converting sunlight into electricity using photovoltaic (PV) modules or through collecting solar heat which it is transformed into electricity (solar thermal electricity).

Solar thermal electricity (STE) generates power without GHG emissions, so it could be a key technology for mitigating climate change. Furthermore, STE plants have an inherent capacity to store thermal energy for subsequent conversion to electricity, even when clouds block the sun, or after shutdown or when power demand increases.

In addition, STE technology shows significant potential for supplying specialised demands such as process heat for industry; co-generation of heating, cooling and power; and water desalination. It is also possible to produce concentrating solar fuels, such as hydrogen, that could help decarbonise the transport and other end-use applications by producing cleaner liquid fuels. Therefore, it is expected that 11 % of global electricity will be generated through solar thermal sources by 2050 (Kasem et al. 2016).

References

Cartelle JJ, Lara M, de la Cruz MP, del Caño A (2015) Assessing the global sustainability of different electricity generation. Energy 89:473–489. doi:10.1016/j.energy.2015.05.110

International Energy Agency (IEA-a) (2015) Energy matters: How COP21 can shift the energy sector onto a low-carbon path that supports economic growth and energy access. https://www.iea.org/publications/freepublications/publication/energy-matters-.html. Accessed 20 May 2016

International Energy Agency (IEA-b) (2015) Special report on energy and climate change. http://www.iea.org/publications/freepublications/publication/weo-2015-special-report-energy-climate-change.html. Accessed 20 May 2016

International Energy Agency (IEA-c) (2015) World energy outlook 2015: executive summary. http://www.iea.org/publications/freepublications/publication/weo-2015.html. Accessed 20 May 2016

International Energy Agency (IEA-d) (2014) Technology roadmap: solar thermal electricity. https://www.iea.org/publications/freepublications/publication/technologyroadmapsolarthermal electricity_2014edition.pdf. Accessed 20 May 2016

Kassem A, Al-Haddad K, Komljenovic D, Schiffauerova A (2016) A value tree for identification of evaluation criteria for solar thermal power technologies in developing countries. Sustain Energ Technol Assessment 16:18–32. doi:10.1016/j.seta.2016.02.003

Chapter 2
Concentrating Solar Thermal Technologies

Concentrating solar thermal technologies belong to an engineering field which can significantly contribute to the delivery of clean, sustainable energy worldwide. This chapter describes the technologies used in the production of solar thermal electricity and process heat divided into medium-concentration solar technology, high-concentration solar technology and the one devoted to solar fuels and industrial processes at high temperatures.

2.1 Introduction

Concentrating solar thermal (CST) technologies can produce electricity on demand when deployed with thermal energy storage, providing a dispatchable source of renewable energy. Therefore, solar thermal electricity (STE) can be provided by smoothing the variability of the energy resource and taking advantage of peak power prices (IEA 2014).

Furthermore, CST technologies can be applied in industrial processes to desalinise water, improve water electrolysis for hydrogen production, generate heat for combined heat and power applications, and support enhanced oil recovery operations. The use of these technologies in a wide range of applications encourages the improvement of their efficiency, which depends on the direct-beam irradiation.

Consequently, arid and semi-arid areas with clear skies are desirable locations where STE plants are installed. In these facilities, curved mirrors are used to concentrate solar radiation onto a receiver which is heated by the radiation. The heat absorbed is transferred to a fluid that passes through the receiver. In order to obtain concentrating solar thermal power, the heated fluid drives a turbine that converts solar heat into electricity (Roldán et al. 2015).

M.I. Roldán Serrano, *Concentrating Solar Thermal Technologies*,
Green Energy and Technology, DOI 10.1007/978-3-319-45883-0_2

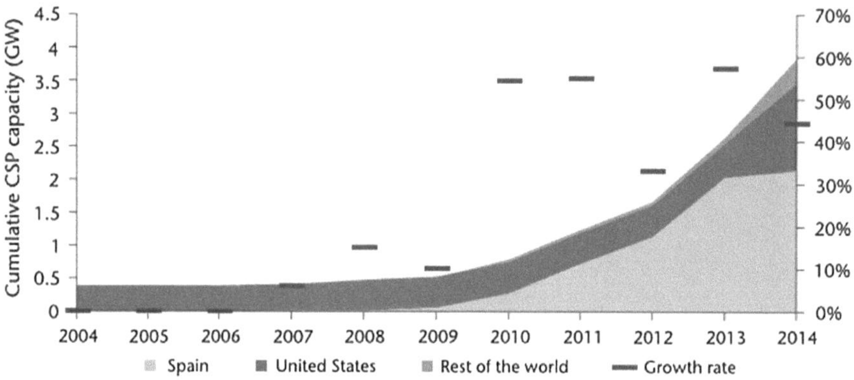

Fig. 2.1 Global cumulative growth of STE capacity (IEA 2014)

The STE industry has experienced robust growth since 2009 which has been most notable in Spain and the United States, and initiated in many other countries with an increasing energy demand (Fig. 2.1, IEA 2014). The largest plants considering the countries with initial STE development are located in United Arab Emirates and India, but others are under construction in Morocco and South Africa. Smaller solar fields, often integrated in larger fossil fuel plants, also can be found in Algeria, Australia, Egypt, Italy and Iran. Furthermore, market prices seem to be falling because new technologies have reached commercial maturity and new concepts have emerged.

2.2 Concentration of Solar Radiation

Solar energy is the most abundant energy resource on earth and the solar radiation reaching the earth's surface equals about 1 kilowatt per square metre (kW/m^2) under clear conditions when the sun is near the zenith. It is comprised of two components: direct or beam radiation, which comes directly from the sun's disk; and diffuse radiation, which reaches the earth after being scattered in all directions by the atmosphere. Hence, global solar radiation is the sum of both components (direct and diffuse radiation).

Many applications require energy at higher temperatures than those reached from incident solar radiation onto the earth's surface. With the aim of achieving high temperatures, solar energy is concentrated in collectors that capture and focus the solar radiation onto a smaller receiving surface.

The relevant measure for the sunlight concentration is the direct normal irradiance (DNI) that corresponds to the density of the available solar resource per unit area on surfaces perpendicular to the direct sunbeam. In humid equatorial places, the atmosphere scatters the sun's rays and DNI is much more affected by clouds and aerosols than global irradiance. The quality of DNI is important in STE plants because the thermal losses of the receiver and the parasitic consumption of the electric auxiliaries are almost constant, regardless of the incoming solar flux. Thus, below a certain level of daily DNI, the net output is null.

High DNI is found in hot and dry regions with reliably clear skies and low aerosol optical depths (subtropical latitudes from 15° to 40° north or south). Closer to the equator, the atmosphere is usually too cloudy. At higher elevations, DNI is also significantly greater, where absorption and scattering of sunlight due to aerosols can be much lower. Thus, the most favourable areas to site STE plants are North Africa, southern Africa, the Middle East, north-western India, south-western United States, northern Mexico, Peru, Chile, the western area of China, Australia, the extreme south of Europe, Turkey, central Asian countries, some places in Brazil, and Argentina (IEA 2014).

As mentioned, the incoming solar radiation can be concentrated in solar collectors whose concentration ratio is evaluated by

$$C = \frac{A_a}{A_r} \tag{2.1}$$

where A_a (m^2) is the aperture area of the concentrator and A_r (m^2) is the receiver area. When the second law of thermodynamics is applied to radiative heat exchange between the sun and the receiver, it is obtained the maximum concentration ratio. Considering a circular concentrator with area A_a, a receiver area A_r and viewing the sun of radius r at distance R, the half-angle subtended by the sun is θ_s. For a perfect concentrator, the radiation from the sun on the concentrator is the fraction of the solar radiation which is intercepted by its aperture. Assuming that the sun is a blackbody at T_s, the heat transferred to the receiver is expressed by the following equation

$$Q_{s \to r} = A_a \cdot \frac{r^2}{R^2} \cdot \sigma \cdot T_s^4 \tag{2.2}$$

where σ is Stephan-Boltzmann's constant (Duffie and Beckman 1980; Roldán et al. 2014).

Similarly, the heat transferred from a perfect receiver at T_r to the sun is given by

$$Q_{r \to s} = A_r \cdot \sigma \cdot T_r^4 \cdot E_{r-s} \tag{2.3}$$

where E_{r-s} is the fraction of radiated energy which reaches the sun. Thus, when $T_s = T_r$, the second law of thermodynamics implies $Q_{s \to r} = Q_{r \to s}$ and the concentration ratio can be evaluated by

$$\frac{A_a}{A_r} = \frac{R^2}{r^2} \cdot E_{r-s} \tag{2.4}$$

Since the maximum value of E_{r-s} is unity, the maximum concentration ratio for circular concentrator is

$$\left(\frac{A_a}{A_r}\right)_{circular,\max} = \frac{R^2}{r^2} = \frac{1}{\sin^2\theta_s} \tag{2.5}$$

The same procedure for linear concentrators leads to

$$\left(\frac{A_a}{A_r}\right)_{linear,\max} = \frac{1}{\sin\theta_s} \tag{2.6}$$

As a result, with $\theta_s = 0.27°$, the maximum possible concentration ratio for circular concentrators is 45,000, and the maximum for linear one is 212. Hence, the higher the temperature delivered, the higher the concentration ratio and the more precise the optics of both the concentrator and the orientation system (Duffie and Beckman 1980).

Concentrators are usually continuous or faceted parabolic reflectors, where the incoming solar beams, parallel to the optical axis, are focused on a point (focus). These reflectors require a solar-tracking mechanism, which can consider one or two axis. The continuous solar-tracking system along one or two axes with a solar concentration on one axis (focal line) reaches medium concentration (523–723 K). The solar-tracking system along two axes with focal point achieves high-concentration levels (temperatures greater than 723 K) (Cabrera et al. 2006).

The combination of different concentrators and receivers makes possible to develop several types of concentrating solar systems divided into: parabolic trough (PT), linear Fresnel reflector (LF), central receiver or solar tower (ST), parabolic dish (PD), and solar furnace as test facility.

2.3 Concentrating Solar Thermal (CST) Technologies

STE plants are gaining in popularity with advances in technology. There is a variety of concentrating solar thermal technologies available nowadays, being solar thermal collectors the major component of solar power systems. As previously stated, these collectors receive the incoming radiation and concentrate solar rays to heat a fluid, which then directly or indirectly drives a turbine and an electricity generator. The concentration of sunlight allows the fluid to reach working temperatures high enough to ensure affordable efficiency in turning the heat into electricity, while limiting heat losses in the receiver.

The four main commercial CST technologies are distinguished by the way they focus the sun's rays and the technology used to receive the solar energy (Fig. 2.2): parabolic-trough collector (PT), solar tower (ST), linear Fresnel (LF) and parabolic dish (PD). They can be classified according to the focus type (line focus or point

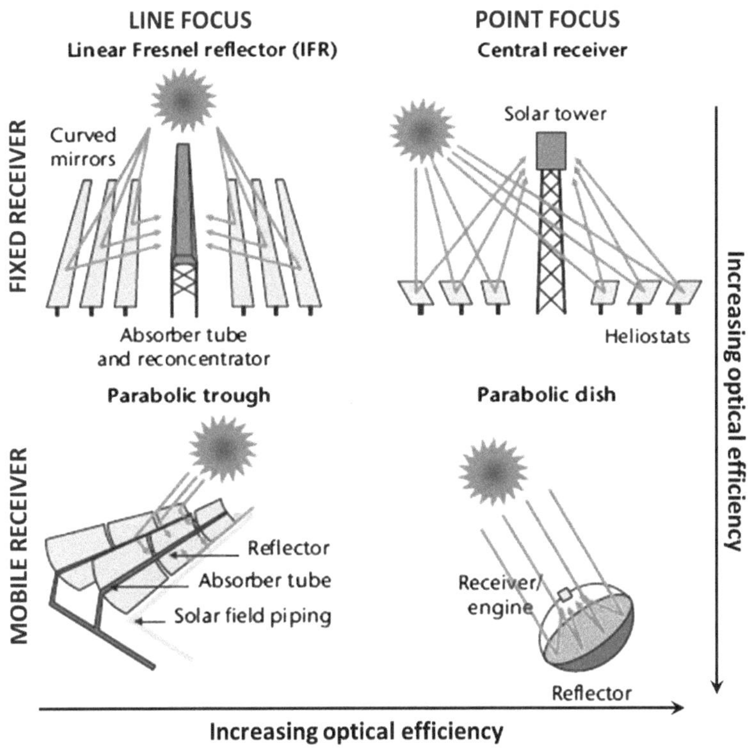

Fig. 2.2 STE technologies (IEA 2014)

one), depending on the receiver type (fixed or mobile) or considering the concentration level (medium or high concentration).

In solar tower and linear Fresnel, the receiver remains stationary and mechanically independent from the concentration system, which is common for all the mirrors. However, the receiver and concentration system move together in PT and PD technologies, enabling an optimal arrangement between concentrator and receiver (IEA 2014; Roldán et al. 2015).

The temperature reached on the receiver is related to the concentration ratio of the collector (Fig. 2.3). Thus, PT and LF reflect the solar rays on a focal line with concentration factors on the order of 60–80 (medium-concentration technologies) and maximum achievable temperatures of about 550 °C. In PD and ST plants, mirrors concentrate the sunlight on a single focal point with higher concentration factors and operating temperatures (high-concentration technologies). On one hand, central receivers achieve a concentration ratio of around 600 and temperatures of 800 °C, and, on the other hand, parabolic dishes reach concentration ratios greater than 1000 that lead to temperatures of 1600 °C. Furthermore, solar furnace is used as test facility able to concentrate around 10,000 times the sunlight reaching temperatures above 2000 °C.

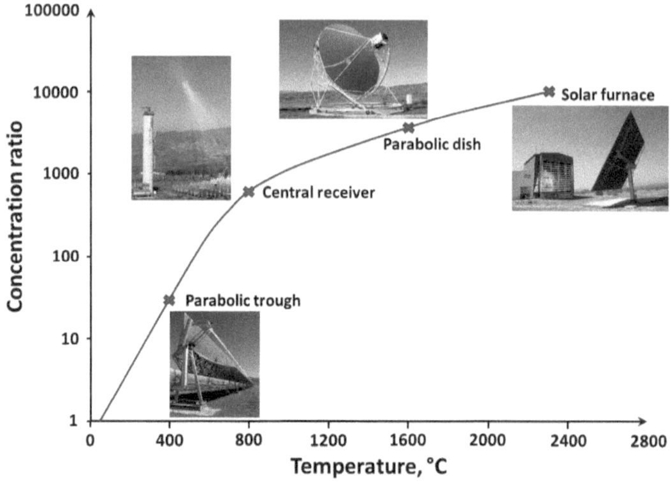

Fig. 2.3 Concentration ratio of the STE technologies

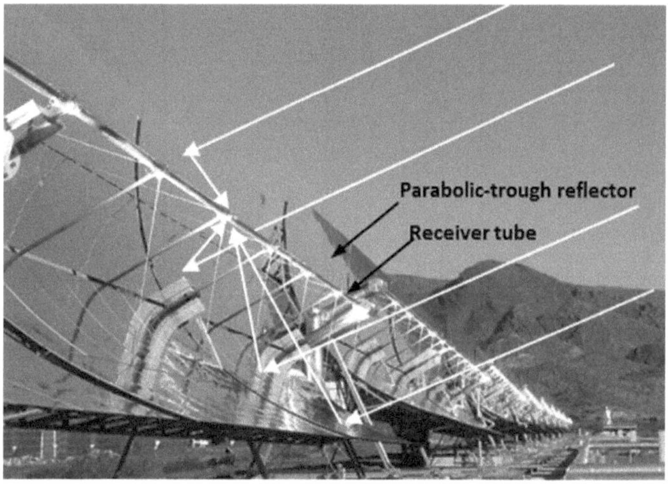

Fig. 2.4 Parabolic trough collector (Roldán et al. 2015)

2.3.1 *Parabolic Trough Collectors*

This is the most mature CST technology, accounting for more than 90 % of the currently installed STE capacity. As illustrated in Fig. 2.4, solar fields using trough systems utilise curved highly-reflective mirrors which focus sunlight onto a linear receiver attached to the focal axis of its parabola. The collectors are connected together in long lines of up to 100 m and track the sun's path throughout the day along a single axis (usually East to West).

Fig. 2.5 Linear Fresnel (Roldán et al. 2015)

The parabolic mirrors are made by bending a sheet of reflective material (silvered low-iron float glass) into a parabolic shape and send the solar beam onto the receiver or absorber tube filled with a specific heat transfer fluid. These metal tubes or pipes have a special coating to maximise energy absorption and minimise infrared re-irradiation. In order to avoid convection heat losses, the tubes work in an evacuated glass envelope.

The thermal energy is removed by the heat transfer fluid (e.g. synthetic oil, molten salt) flowing in the heat-absorbing pipe and transferred to a steam generator to produce the super-heated steam that drives the turbine. Once the fluid transfers its heat (temperatures of up to 400 °C), it is recirculated into the system for reuse. The steam is also cooled, condensed and reused. Furthermore, the heated fluid in PT technology can also provide heat to thermal storage systems, which can be used to generate electricity at times when the sun is not shining.

Most PT plants currently in operation have capacities between 14 and 80 MW_e, efficiencies of around 14–16 % (i.e. the ratio of solar irradiance power to net electric output) and maximum operating temperatures of 390 °C, which is limited by the degradation of synthetic oil used for heat transfer. The use of molten salt at 550 °C for heat transfer purposes in PT plants is under investigation. High temperature molten salt may increase both plant efficiency (e.g. 15–17 %) and thermal storage capacity (Roldán et al. 2015).

2.3.2 *Linear Fresnel*

Linear Fresnel (LF) is similar to PT collector, with slight differences (Fig. 2.5). It uses a series of ground-based, flat or slightly curved mirrors placed at different angles to concentrate the sunlight onto a fixed receiver located several meters above the mirror field. Each line of mirrors is equipped with a single axis tracking system to concentrate the sunlight onto the receiver which consists of a long, selectively-coated tube. The facility usually uses water as heat transfer fluid, which passes through the receiver and it is converted into steam (DSG or Direct Steam Generation). Since the focal line in the LF plant can be distorted by astigmatism, a secondary mirror is placed above the receiver to refocus the sun's rays. Alternatively, multi-tube receivers can be used to capture sunlight with no secondary mirror (Roldán et al. 2015).

Flat mirrors and shared receivers result in lower expenses, while at the same time, this technology benefits from the long-term success. Furthermore, similar to the PT system, linear Fresnel does not need two-axis tracking since the sun will be focused on a part of the system throughout the year.

The main advantages of LF compared to PT systems are the lower cost of ground-based mirrors and solar collectors (including structural supports and assembly). While the optical efficiency of the LF system is lower than that of the PT systems (i.e. higher optical losses), the relative simplicity of the plant translates into lower manufacturing and installation costs compared to PT plants.

Thus, the mirror aperture can be augmented more easily than with troughs, and secondary reflection makes possible higher concentration factors, decreasing thermal losses. However, LF has greater optical losses than troughs when the sun is low in the sky. This reduces power generation in early morning and late afternoons, and also in winter, but can be overcome in part by the use of higher operating temperatures than PT plants (IEA 2014).

Therefore, it is not clear whether LF electricity is cheaper than the one from PT plants. Furthermore, as LF systems use direct steam generation, thermal energy storage is likely to be more challenging and expensive. Thus, LF is the most recent CST technology with only a few plants in operation. The largest solar thermal plant using LF technology is Puerto Errado in Spain with a capacity of 30 MW_e (Kassem et al. 2016).

2.3.3 *Solar Tower*

In the ST plants (Fig. 2.6), also called central receiver systems (CRS) or power tower, a large number of computer-assisted mirrors (heliostats) track the sun individually over two axes. Heliostats are less expensive than trough mirrors because they utilise standard flare glass, instead of glass that is manufactured at specific curves. They concentrate the solar radiation onto a single receiver at the top of a central tower where the solar heat drives a thermodynamic cycle and generates electricity. ST plants can achieve higher temperatures than PT and LF systems because they have higher concentration factors (Fig. 2.3). The CRS can use water-steam (DSG), synthetic oil or molten salt as the primary heat transfer fluid.

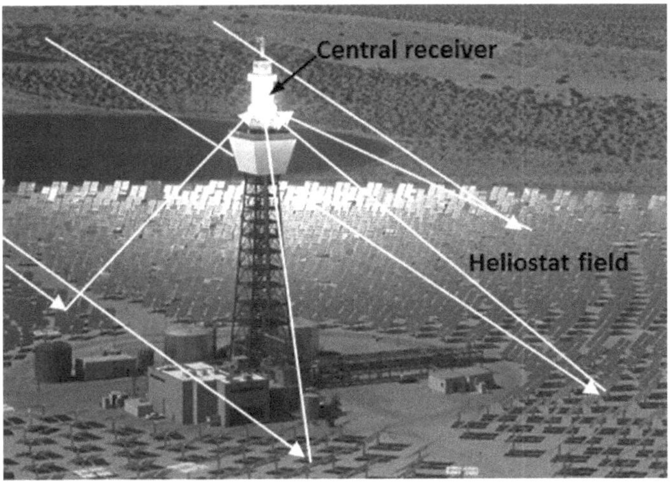

Fig. 2.6 Solar tower (Roldán et al. 2015)

The use of high-temperature gas is also being considered (e.g. atmospheric air in volumetric receivers).

In a direct steam ST, water is pumped up the tower to the receiver, where concentrated thermal energy heats it to around 550 °C. The hot steam then powers a conventional steam turbine. When DSG is used as heat transfer fluid, it is not required a heat exchanger between the primary transfer fluid and the steam cycle, but the thermal storage is more difficult.

Depending on the primary heat transfer fluid and the receiver design, maximum operating temperatures may range from 250 to 300 °C (using water-saturated steam) to 390 °C (using synthetic oil) and up to 565 °C (using molten salt and water-superheated steam). Temperatures above 800 °C can be obtained using gases (e.g. atmospheric air). Thus, the temperature level of the primary heat transfer fluid determines the operating conditions (i.e. subcritical, supercritical or ultra-supercritical) of the steam cycle in the conventional part of the power plant.

ST plants can be equipped with thermal storage systems whose operating temperatures also depend on the primary heat transfer fluid. Today's best performance is obtained using molten salt at 565 °C for both heat transfer and storage purposes. This enables efficient and cheap heat storage and the use of efficient supercritical steam cycles (Roldán et al. 2015).

High-temperature ST plants offer potential advantages over other CST technologies in terms of efficiency, heat storage, performance, capacity factors and costs. In the long run, they could provide the cheapest STE, but more commercial experience is needed to confirm these expectations. However, a large ST plant can require thousands of computer-controlled heliostats, that move to maintain point focus with the central tower from dawn to dusk, and they typically constitute about 50 % of the plant's cost.

The largest solar thermal plant operating through power towers is the Ivanpah Solar Power Facility in the USA, with a capacity of 392 MW_e. The plant gathers three distinct towers, each with its own turbine, based on DSG technology and no storage. There are two more facilities in Spain, each with approximately 20 MW_e of capacity, and several other facilities with lower capacities in Turkey, India, and other countries (Kassem et al. 2016).

Larger ST plants have expansive solar fields with a high number of heliostats and a greater distance between them and the central receiver. This results in more optical losses, atmospheric absorption and angular deviation due to mirror and sun-tracking imperfections. Therefore, ST still has room for improvement of its technology.

2.3.4 Parabolic Dish

Parabolic dish (PD) systems (Fig. 2.7) consist of a concave dish shaped concentrator that reflects sunlight into a receiver placed at the focal point of the dish. The receiver may be a Stirling engine or a micro-turbine. PD requires two-axis sun tracking system to follow the sun from east to west during the day, and from north to south throughout the year. This technology offers very high concentration factors and operating temperatures (Fig. 2.3).

To date, there are no large utilities using PD technology, due to several difficulties. The design of reliable engines for large plants is still under development. In addition, the initial cost of such systems is high in comparison with the CST

Fig. 2.7 Parabolic dish (Roldán et al. 2015)

technologies previously described, and there are also challenges related to the storage capability. Nevertheless, the Stirling dish system has the highest efficiency in the conversion from heat to electricity, with a net average annual yield rate that is 18–23 % higher than any other solar energy system.

Therefore, the main advantages of PD systems include high efficiency (i.e. up to 30 %) and modularity (i.e. 3–50 kW), which is suitable for distributed generation. Unlike other STE options, PD systems do not need cooling systems for the exhaust heat. This makes PD suitable for use in water-constrained regions, though at relatively high electricity generation costs compared to other CST technologies.

PD technology is currently considered a potential technology for STE generation and many pilot projects have been launched in the USA and Spain. However, the PD system is still under demonstration and investment costs are still high. Thus, with more research and development, it could be a potential alternative candidate technology for STE plants (Roldán et al. 2015; Kassem et al. 2016).

2.3.5 Solar Furnace

Solar furnaces reach the highest energy levels in concentrating solar systems (over 10,000 kW/m^2, see Fig. 2.3). Therefore, they are used as a test setup for high-temperature processes and other applications, such as material treatment, development and investigation of new solar receivers, and simulation of thermal effects from highly concentrated heat flux, among others.

The facility (Fig. 2.8) consists of an optical system with one or more heliostats, which reflect the solar radiation onto a concentrator. This reflector can be composed by a parabolic mirror or a group of spherical mirrors. The furnace power can be

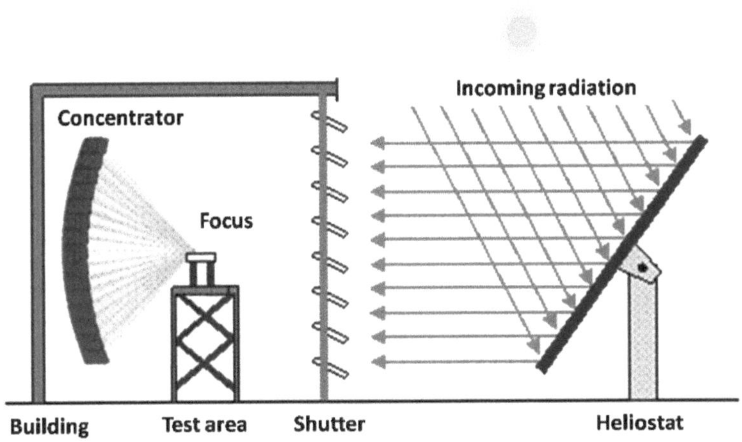

Fig. 2.8 Solar furnace (Roldán et al. 2014)

attenuated by a shutter, which control the amount of solar radiation received onto the concentrator. The concentrated radiation reaches the test area, which is located at the concentrator focus.

2.4 Status of CST Technologies

2.4.1 Medium-Concentration Solar Technology

Medium-concentration solar power plants use the line focusing parabolic solar collector at a temperature of about 400 °C. Significant advances have been made in parabolic collector technology together with organic Rankine cycle technology to improve the performance of parabolic trough STE plant. Furthermore, the traditional sun-tracking unit with sensors that detect the position of the sun has been replaced by a system based on the calculation of the sun position using a mathematical algorithm.

The Solar Energy Generating System in the USA is the largest parabolic trough power plant complex in the world, with a capacity of 354 MW_e. A recent development in cost effective concentrators is the design of Euro Trough, a new parabolic trough concentrator, in which an advanced lightweight structure is used to achieve cost efficient solar power generation. Parabolic trough STE plants can collect up to 70 % of the incident solar radiation and achieve a peak electrical conversion efficiency of 20–25 % (Siva Reddy et al. 2013) whose improvement is the main challenge for this technology.

Linear Fresnel facility is similar to a PT with the advantages of low costs for structural support and reflectors, fixed fluid joints, a receiver separated from the reflector system, and long focal lengths that allow the use of flat mirrors. While in 2010 only a couple of prototypes using LF reflectors were operating, a 30 MW_e LF plant built in Calasparra (Spain) started up in early 2012, and a 125 MW_e commercial located in India began operating in 2014. All LF plants currently use DSG and they do not have thermal storage, which is a challenging development issue for this technology.

2.4.2 High-Concentration Solar Technology

The parabolic dish-Stirling engine and the central tower receiver are primarily tried for high-temperature solar thermal power plants. Stirling-dish STE plants developed for commercial applications generate power for its supply in isolated communities and villages of rural areas. Furthermore, parabolic dish has a complete two axes tracking of the concentrator aperture that would increase the amount of the incoming radiation by avoiding the cosine effect. At the focal point, the Stirling receiver absorbs solar radiation and transfers the thermal energy to the engine. The

Table 2.1 Performance data for examples of solar tower plants with different working fluid

Parameter	Solar 100 plant	PS-10 plant
Working fluid	Molten salt	Air
Plant rating	100	10
Annual solar insolation (kWh/m^2)	2700	2063
Field area (m^2)	1,466,000	89,271
Receiver thermal rating (MW)	796	55
Thermal storage size (MWh)	3820	–
Steam generator rating (MW)	254	5.34
Annual net energy production (MWh)	613,000	19,200
Peak net efficiency	0.22	0.17
Annual net efficiency	0.16	0.12

Source Siva Reddy et al. (2013)

main Stirling absorbers are typically direct irradiated receivers, heat pipe receivers, and volumetric ones.

The heat pipe absorbers vaporise a liquid metal such as sodium on the absorber surface and the gas condenses on the Stirling engine heater tubes to transfer the energy to the working fluid. Heat pipe receivers reach more uniform temperature distribution on the tubes, resulting in longer life for both the absorber and engine heater head in comparison with the direct irradiated absorber. Volumetric receivers are potentially more cost effective and reliable than the heat pipe receivers, but the design of reliable engines for large PD plants is still under development.

In central receiver systems, there are different receiver types depending on their configuration and the heat transfer medium. The configuration can be either external or cavity type. In a cavity receiver, the radiation reflected from the heliostats passes through the aperture into a box-like structure before impinging on the heat transfer surface. External receivers can be designed with a flat-plate or cylindrically shaped tubular panels.

The performance data on ST plants for two different receivers and heat transfer media (molten salt and air) are collected in Table 2.1.

2.4.3 Performance of CST Technologies

More than 90 % of the installed STE capacity in 2014 consisted of PT plants; ST plants total about 170 MW and LF plants of around 40 MW. A comparison of CST technology performance is shown in Table 2.2.

According to the previous data, the facility performance could be improved by considering an optimised receiver design and operating conditions, defining the best operation strategy for each facility and studying alternative working fluids which drive the turbine in more efficient power blocks. These should be the main issues addressed in future developments of CST technologies.

Table 2.2 Performance of CST technologies

	PT	PT	PT	ST	ST	ST	LF	PD
Storage	No	Yes	Yes	No/yes	No/yes	Yes	No	No
Status	Comm	Comm	Demo	Demo	Comm	Demo	Demo	Demo
Capacity (MW)	15–80	50–280	5	10–20	50–370	20	5–30	0.025
HTF	Oil	Oil	Salt	Steam	Steam	Salt	Sat.st	Na
HTF temperature (°C)	390	390	550	250	565	565	250	750
Storage fluid	No	Salt	Salt	Steam	Na	Salt	No	No
Storage time (h)	0	7	6–8	0.5–1	Na	15	0	0
Storage temperature (°C)	Na	380	550	250	Na	550	Na	Na
Efficiency (%)	14	14	14/16	14	16	15/19	11/13	25/30
Cap. Factor (%)	25–28	29–43	29–43	25–28	25–28	55–70	22–24	25–28
Optical efficiency	H	H	H	M	M	H	L	VH
Concentration	70–80	70–80	70–80	1000	1000	1000	60–70	>1300
Land (ha/MW)	2	3	2	2	2	2	2	Na
Cycle	Sh.st	Sh.st	Sh.st	Sat.st	Sh.st	Sh.st	Sat.st	Na
Cycle temperature (°C)	380	380	540	250	540	540	250	Na
Grid	On	On	On	On	On	On	On	On/off

HFT heat transfer fluid; *Sat.st* satured steam; *Sh.st* superheated steam; *L* low; *M* middle; *H* high; *VH* very high; *Na* not applicable
Source Roldán et al. (2015)

References

Cabrera JA, Cuesta MJ, Pérez M (2006) Concentrating solar power: current state and sector actors. CIEMAT Tech Report

Duffie JA, Beckman WA (1980) Solar engineering of thermal processes. Wiley, New York

International Energy Agency (2014) Technology roadmap: solar thermal electricity. https://www.iea.org/publications/freepublications/publication/technologyroadmapsolarthermalelectricity_2014edition.pdf. Accessed 20 May 2016

Kassem A, Al-Haddad K, Komljenovic D, Schiffauerova A (2016) A value tree for identification of evaluation criteria for solar thermal power technologies in developing countries. Sustain Energ Technol Assessment 16:18–32. doi:10.1016/j.seta.2016.02.003

Roldán MI, Valenzuela L, Fernández J (2014) Computational fluid dynamics in concentrating solar technologies. In: Al-Baghdadi MARS (ed) Computational fluid dynamics applications in green design, 1st edn. International Energy and Environment Foundation, Iraq

Roldán MI, Fernández J, Valenzuela L, Vidal A, Zarza E (2015) CFD Modelling in solar thermal engineering. In: Al-Baghdadi MARS (ed) Engineering applications of computational fluid dynamics: volume 3, 1st edn. International Energy and Environment Foundation, Iraq

Siva Reddy V, Kaushik SC, Ranjan KR, Tyagi SK (2013) State-of-the-art of solar thermal power plants—A review. Renew Sust Energ Rev 27:258–273. doi:10.1016/j.rser.2013.06.037

Chapter 3
Computational Fluid Dynamics in Green Design

Computational Fluid Dynamics simulation is one of the most appropriate methods to analyse different phenomena involving fluid flows. It is used to reproduce the operating conditions of a facility and to analyse the parameters which determine its performance. Thus, CFD simulation can be used in the design and optimisation of concentrating solar thermal technologies that belong to the field of green engineering in which environmental and human health impacts are minimised. This chapter describes the advantages of using CFD for the design and optimisation of CST plants as an example of green engineering.

3.1 Introduction

Computational Fluid Dynamics (CFD) is the science of predicting fluid flow, heat transfer, mass transfer, phase change, chemical reaction, mechanical movement, stress or deformation of related solid structures, and related phenomena by solving the mathematical equations that govern these processes using a numerical algorithm on a computer. The results of CFD analyses are relevant in conceptual studies of new designs, detailed product development, facility troubleshooting, optimisation of the operating conditions, and redesign.

There is a pressing need to accelerate the development of advanced energy technologies in order to address the global challenges of clean energy, climate change and sustainable development. This is the basis of the emerging concept called "green design". It can be defined as the design, commercialisation and use of processes and products that are feasible and economical while reducing the generation of pollution at the source and minimising the risk to human health and the environment (Al-Baghdadi 2014).

M.I. Roldán Serrano, *Concentrating Solar Thermal Technologies*,
Green Energy and Technology, DOI 10.1007/978-3-319-45883-0_3

Related concepts are sustainable or environmental design. Therefore, in the engineering field, the term "green" is referred to process and product generation in a way that minimises pollution, promotes sustainability, and protects human health without sacrificing economic viability and efficiency. More widely, the term has become associated with sustainable development, in which processes and products can continue to be produced indefinitely with a minimum of resource depletion or environmental degradation.

Concentrating solar thermal (CST) technologies belong to the green engineering field, due to the utilisation of a renewable energy source (solar energy) as alternative to fossil fuels and their positive contribution to sustainable development and processes without greenhouse gas emissions. The analysis and prediction of the thermal and fluid-dynamic behaviour in CST plants is essential in order to enhance the thermal performance of the facility. For that purpose, CFD analysis complements testing and experimentation, by reducing the total effort required in the experiment design and data acquisition.

CFD reinforces physical modelling and other experimental techniques by providing a detailed look into the fluid flow patterns in the facility, including complex physical processes such as turbulence, chemical reactions, heat and mass transfer, and multiphase flows. In many cases, virtual models can be built and analysed consuming a fraction of the time and cost required by experimental procedures. This makes it possible to investigate more design options and scenarios under extreme conditions. Moreover, CFD modelling provides insights into the fluid flow issues that would be too costly or simply prohibitive by experimental techniques. The added insight and understanding gained from flow modelling gives confidence in the design proposals, avoiding the added costs of over-sizing and over-specification, while reducing risk and increasing the facility life-time.

Therefore, the use of Computational Fluid Dynamics to simulate engineering phenomena continues to grow throughout many engineering disciplines. CFD provides a reliable tool to assist engineers and researchers in the design of industrial equipment or innovative facilities often reducing or eliminating the need for performing trial-and-error experimentation.

3.2 Green Engineering

Green engineering considers that decisions to protect human health and the environment can have the greatest impact and cost-effectiveness when they are applied in the design and development phase of a process or product.

Therefore, green engineering processes and products are based on the following principles (EPA 2015):

- The holistically use of system analysis and the integration of environmental impact assessment tools.
- Conservation and improvement of natural ecosystems while protecting human health and well-being.
- The use of life-cycle thinking in all engineering activities.
- To ensure that all material and energy inputs and outputs are as inherently safe and benign as possible.
- Minimisation of the natural resource depletion.
- To avoid waste and greenhouse gas emissions.

Additionally, green engineering develops and applies engineering solutions adapted to the area where the facility is located. These solutions should improve and obtain an innovative technology to achieve sustainability.

In this context, CST technologies require complete and efficient green engineering in order to obtain the maximum performance of each facility. The thermosolar field is still emerging, and, in many cases, the technology and facilities used are experimental. Therefore, it is necessary to apply advanced simulation tools to predict the behaviour of the heat transfer fluid in the solar thermal installation and to define and optimise the operating conditions of the system.

3.3 Computational Fluid Dynamics (CFD)

Analysis and control of technical and scientific processes are essential to reach the appropriate facility operation and the optimisation of the process developed. For this purpose, it is required a research work that can be addressed by experimental procedure and/or by theoretical simulation. Experimental procedure obtains information by observing the physical phenomena in themselves, and theoretical simulation studies these phenomena by computer programme based on mathematical models. These models in turn come from scientific theories which explain real phenomena. Therefore, the strategy selected to analyse the process should be based on the accuracy obtained, cost required and feasibility of the method used (Ueberhuber 1997).

The rapid development of computer science and numerical techniques has encouraged the study of real phenomena by computer modelling. Computational Fluid Dynamics (CFD) is a field of Fluid Mechanics that analyses systems with fluid flows, heat transfer and other associated phenomena, such as chemical reactions, by means of computer simulation and proper numerical algorithms. Thus, CFD solves the flow governing equations (i.e. the fundamental conservation equations in mass, momentum, and energy) to predict quantitatively the fluid behaviour, often associated with other phenomena:

- simultaneous heat flow
- mass transfer (e.g. perspiration, dissolution)
- phase change (e.g. melting, freezing, boiling)
- chemical reaction (e.g. combustion, rusting)
- mechanical movement (e.g. pistons, fans, rudders)
- stresses in and displacement of immersed or surrounding solids.

The validated CFD models can be used to understand physics, improve designs, optimise systems and operating conditions, guide procedures, and influence decision-makings.

3.3.1 Advantages of CFD Simulation

Unlike experimental procedure, the theoretical prediction of a phenomenon consists of a mathematical model coming from a previous physical model. The mathematical formulation is solved by numerical methods able to simplify the differential equations of the model. Hence, the main advantages of the theoretical calculations in comparison with the experimental research are the following (Patankar 1980):

- Relatively low cost due to the fast development in computer industry. As a consequence, computational modelling becomes less expensive in investment and can provide more informative results within much shorter time. The costs are likely to continue decreasing with more powerful rather cheap computers and more users of CFD software. For most studies, the cost of CFD is almost negligible in comparison with the experimental approach, whether on-site or mock-up experiments.
- CFD simulations can be executed in a short period of time (ranging from a few seconds to a few days depending on the physics of the problem and the resolution requirements) due in part to the availability of robust and fast computers. Quick turnaround means engineering data can be introduced early in various decision-making processes. This is especially true for design projects.
- CFD provides the ability to theoretically simulate any physical conditions, especially those that cannot be easily tested in experiments. CFD can effectively and safely model the situations under extreme or ideal conditions, such as extreme-hot/cold and high-toxic scenarios, in which the measurement is usually very difficult or even impossible.
- CFD is able to simulate ideal conditions and allows great control over the physical process, and provides the ability to isolate specific phenomena for study. Thus, it is possible to deliberately study the influence of a particular design feature on the whole system performance by adjusting a specific parameter selected while keeping others unchanged in the CFD simulation.

- Comprehensive information can be obtained by CFD modelling because it is able to examine a large number of locations in the domain of interest, and yields a wide set of flow variables (e.g. detailed distributions of air velocity, pressure, temperature, and substance concentrations), mostly under a single computation. The information allows one to have a global knowledge of flow and heat transfer, rather than limited observations based on a few of measurement points.
- Operation easiness is due to the growing attention and development in intelligent graphic user interface technologies. A CFD user can easily change and test different modelling scenarios once the initial model is well built and validated.

3.3.2 Disadvantages of CFD Simulation

The usefulness of the computational analysis depends on the validity of the mathematical model on which it is based, on the proper definition of the parameters and conditions of the process simulated, and on the numerical methods used to solve the model. Therefore, when it is required the construction of several prototypes to analyse the system behaviour, which is time-consuming and costly, the best option is the use of theoretical simulation by numerical methods. However, it presents some issues (Al-Baghdadi 2014; Versteeg and Malalasekera 1995):

- The accuracy of CFD results is strongly dependent on whether the flow governing equations solved by a computer can correctly describe the flow physics. In addition, developing an appropriate computer model of a reality sometimes requires profound knowledge, prior experience, and creativity of handling similar problems. Simplification methods and processes of complex real objects into computer recognizable models will largely influence the accuracy of simulation results.
- The boundary conditions supplied by the user determine the accuracy of CFD predictions. These boundary conditions are crucial for the accuracy of the CFD results, but most of these are based on empirical data or even experienced guess. The circumstance may become more challenging when time-varying boundary conditions are required for an unsteady calculation, in which the dynamic measurement data is usually unavailable and difficult to evaluate.
- Continuous space and time domains in physics must be discretised into separate systems before a computer code can recognise and process. Various numerical schemes and methods are utilised during the process, which may lead to unstable, un-converged, and unrealistic results if handled improperly.
- Since the flow governing equations are highly non-linear and strongly self-coupled, CFD applies numerical methods, such as the finite volume method, to discretise the partial differential equations and obtain the corresponding algebraic equations that can be solved iteratively. The numerical approximation and iterative calculation may introduce various uncertainties and instabilities. In order to avoid these problems, advanced algorithms and techniques were created

to ensure the calculation toward a convergent and stable direction, such as the false-time-step and relaxation factor methods. However, when more sophisticated mathematical models and numerical techniques are developed and applied to handle the complex problems, the numerical stability and convergence issues are always of big concern.

Therefore, when there is an adequate mathematical description to study a phenomenon, the main problems are: high computational cost for complex geometries, difficult calculation for extremely fast and small-scale phenomena, and unclear definition of the real solution when the mathematical model admits more than one solution.

In cases with improper mathematical description, it is additionally found uncertainty problems because numerical results are not in agreement with the reality. In this situation, some experimental backup is highly desirable.

Nevertheless, research in computational methods is aimed at making them more reliable, accurate and efficient. Thus, the disadvantages mentioned will be solved as this research progresses.

3.4 Principles of CFD Simulation

Increasingly, CFD is becoming an essential component in the design of industrial products and processes. It has been developed to provide a capability comparable to other computer-aided engineering tools. The slow progress in CFD codes is due to the tremendous complexity of the underlying behaviour, which makes it difficult to obtain at the same time an economical and exhaustive solution of the phenomenon analysed. However, the availability of affordable high performance computing hardware and the introduction of user-friendly interfaces have led to a significant increase in the CFD use (Roldán 2012).

CFD codes are based on numerical algorithms that can address fluid flow problems. In order to provide an easy solution for the issue analysed, commercial CFD packages include sophisticated user interfaces to input problem parameters and to examine results. Hence, CFD codes consist of three elements: pre-processor, solver, and post-processor.

3.4.1 Pre-processing

Pre-processing is the methodology followed to define the input of a flow problem in a CFD program by an operator-friendly interface and the subsequent transformation of the input into a form suitable for use by the solver. This stage involves different activities (Versteeg and Malalasekera 1995):

- Definition of the solution domain
- Grid generation to divide the computational domain into a number of smaller, non-overlapping sub-domains. The grid consists of cells or control volumes and calculation nodes
- Selection of the physical and chemical phenomena involved
- Definition of fluid properties
- Specification of appropriate boundary conditions.

The nodes inside each cell define the solution to a flow problem and the desirable accuracy of the CFD solution will depend on the number of cells in the grid. In general, the larger the number of cells the better the solution accuracy, but the computational cost increases at the same time. Optimal meshes are frequently non-uniform, considering finer grids in areas where there are large variations of the fluid behaviour and coarser ones in regions with relatively little change. The latest released versions of CFD codes include a self-adaptive meshing capability and they automatically refine the grid in areas of rapid variations. However, the user should check the quality of the mesh obtained by the CFD code, considering cell parameters such as equiangle skew and aspect ratio, among others.

The definition of the geometry and grid generation require over 50 % of the time spent on a CFD project. In order to minimise the execution time of the project, the CFD codes tend to include their own CAD interface and/or tools to import data from external surface modellers and mesh generators. Up-to-date pre-processors also give the user access to libraries of material properties for the most conventional fluids and they allows implementing special physical and chemical process models to reproduce phenomena such as turbulence, radiative heat transfer and combustion.

3.4.2 Solver

In order to solve the fluid dynamics problem, there are three different streams of numerical solution techniques: finite difference, finite element and spectral methods. Numerical methods are the basis of the solver and they consider three main steps (Versteeg and Malalasekera 1995):

- Approach of the unknown flow variables by means of simple functions
- Discretisation by substitution of the approaches into the governing flow equations and subsequent mathematical transformations
- Solution of the algebraic equations.

The previously mentioned techniques differ depending on the way in which the flow variables are approximated and on the discretisation processes. Thus, *finite difference methods* describe the unknown variables of the flow problem by point samples at the nodes of a grid of coordinate lines. Finite difference approaches of derivatives of the flow variable are often generated by truncated Taylor series expansions in terms of point samples of the flow variable at each grid point and its immediate neighbours. The derivatives of the governing equations are replaced by finite differences, providing an algebraic equation for the values of the flow variable at each grid point.

Finite element methods use simple piecewise functions valid on elements to describe the local variations of unknown flow variables. If the piecewise approximating functions for the flow variable are substituted into the governing equation, it will not precisely satisfied by the exact solution and a residual is defined to measure the errors. Residuals are minimised by multiplying them by a set of weighting functions and integrating the resulting expression. Finally, it is obtained a set of algebraic equations for the unknown coefficients of the approximating functions.

Spectral methods approach the unknown flow variable by truncated Fourier series or series of Chebyshev polynomials. In this method, the approaches are not local but valid throughout the entire computational domain. The unknown flow variables are replaced by the truncated series into the governing equation. The constraint that provides the algebraic equations for the coefficients of the Fourier or Chebyshev series is given by a weighted residual concept or by making the approach functions coincide with the exact solution at a number of grid points.

Additionally to the preliminary information given related to the main numerical methods used by CFD codes, it is presented the numerical method used in the CFD simulations performed to study specific flow problems for concentrating solar thermal technologies included in this book. This method is called *finite volume method* and it can be considered as a special finite difference formulation whose numerical algorithm consists of three main steps:

- Formal integration of the governing equations of fluid flow over all the finite control volumes of the solution domain
- Discretisation to convert the integrated equations into a system of algebraic equations. This involves the substitution of a variety of finite-difference-type approaches for the terms in the integrated equation, representing flow processes such as convection, diffusion and sources
- Solution of the algebraic equations by an iterative method.

The control volume integration distinguishes this numerical method from other CFD techniques, because the resulting statements lead to the conservation of relevant properties for each finite size cell. The relationship between the numerical algorithm and the physical conservation principle makes the concept of finite volume method much more simple to understand than finite element and spectral methods. The conservations of a flow variable within a finite control volume can be defined as a balance between the different processes tending to increase or decrease.

The discretisation techniques of the CFD codes allow the treatment of the main transport phenomena (convection and diffusion) as well as the source terms and the rate of change with regard to time. As mentioned, physical phenomena are non-linear and complex. Therefore, an iterative solution approach is required such as SIMPLE algorithm in order to ensure correct coupling between pressure and velocity.

3.4.3 *Post-processing*

In the post-processing, the information required by the user is obtained to understand and study the physical phenomena, including the visualisation of the resulting data. The leading CFD packages are equipped with versatile and robust data visualisation tools, which include:

- Display for domain geometry and mesh
- Line and shaded contour plots
- Vector plots
- 2D and 3D surface plots
- Particle tracking
- Colour postscript output
- View manipulation
- Dynamic result display
- Data export tools.

The validity of the simulation data should be studied in order to demonstrate the accuracy of the CFD model and the reproducibility of the phenomenon. For the validation processes, it is usually required experimental data obtained from reference cases.

3.5 Governing Equations for Fluid Flow

The equations applied to Fluid Mechanics are obtained from the principles of conservation in Mechanics and Thermodynamics considered for a fluid volume. The resulting expressions are called Navier–Stokes equations which describe the motion of fluid substances such as liquids and gases. These equations state that changes in momentum of fluid particles depend only on both the external pressure and internal viscous forces acting on the fluid. These are differential equations describing the balance of forces which act at any given region of the fluid.

This means that, in most cases, solutions of the Navier–Stokes equations for a given physical problem are obtained by numerical methods. In practical terms, only the simplest cases can be solved exactly in this way. These cases generally involve non-turbulent, steady flow in which the Reynolds number is small. For more complex situations, involving turbulence, solutions of the Navier–Stokes equations can currently only be found by a computer (Blazek 2008).

Therefore, the dynamic behaviour of a fluid is defined by the three conservation laws: conservation of mass, momentum, and energy. This means that the continuity (3.1), momentum (3.2) (Bachelor 1967) and energy (3.3) (Versteeg and Malalasekera 1995) equations described for the fluid medium are included in the CFD model:

$$\frac{\partial \rho}{\partial t} + \nabla \cdot \left(\rho \vec{v}\right) = S_m \tag{3.1}$$

$$\frac{\partial}{\partial t}\left(\rho \vec{v}\right) + \nabla \cdot \left(\rho \vec{v}\,\vec{v}\right) = -\nabla p + \nabla \cdot (\bar{\bar{\tau}}) + \rho \vec{g} + \vec{F} \tag{3.2}$$

$$\frac{\partial}{\partial t}(\rho E) + \nabla \cdot \left(\vec{v}(\rho E + p)\right) = \nabla \cdot \left(k_{eff} \nabla T - \sum_j h_j \vec{J_j} + \left(\bar{\bar{\tau}}_{eff} \cdot \vec{v}\right)\right) + S_{he} \tag{3.3}$$

where ρ is the density of the fluid, t is elapsed time, $\vec{v}$ is the velocity vector with respect to the coordinate system, S_m is the mass source, p is the static pressure, $\bar{\bar{\tau}}$ is the stress tensor, $\rho\ \vec{g}$ is the gravitational body force, $\vec{F}$ is the external body force, E is the energy transfer ($E = h - \frac{p}{\rho} + \frac{v^2}{2}$), k_{eff} is the effective conductivity which includes the turbulence thermal conductivity, h_j is the enthalpy of species j, $\vec{J_j}$ is the diffusion flux of species j, $\bar{\bar{\tau}}_{eff}$ is the viscous stress tensor, and S_{he} is the volumetric heat source.

The equations previously described are the basis of a CFD model, which will include additional ones (radiative and viscosity models, among others) depending on the complexity of the problem studied.

3.6 Solving by CFD

As mentioned, to solve fluid flow problems, it is important to be aware of the complexity of the physics involved and of the result reliability which depends on the physic implemented in the model and on the user skills. Prior to setting up and running a CFD simulation, it is necessary to identify and formulate the flow problem regarding the physical and chemical phenomena related to the case selected.

Some of the decisions that have to be taken in the model definition are linked to the selection of a 2D geometry or a 3D one, the assumption of excluding the effects of ambient temperature or pressure variations on the density of an air flow, solving the turbulent flow equations or neglecting the effects of small bubbles dissolved in tap water (two-phase flow). It is necessary to make assumptions even in the simplest flow issues to minimise the complexity but always providing reliable results which depend largely on the modelling skills of the user.

Furthermore, a good understanding of the numerical solution algorithm is also essential and the appropriateness of their solution can be evaluated by mathematical concepts such as convergence, consistency and stability. These properties are analysed by some authors (Versteeg and Malalasekera 1995).

- *Convergence* is the property of a numerical method to produce a solution which approaches the exact solution as the grid spacing; control volume size is reduced to zero. More formally, in mathematics, convergence describes limiting behaviour, particularly of an infinite sequence or series toward some limit. To assert convergence is to claim the existence of a limit, which may be unknown.
  As this definition indicates, the exact solution to the iterative problem is unknown, but it is important to be sufficiently close to the solution for a particular required level of accuracy. Convergence therefore does need to be associated with a requirement for a particular level of accuracy. This requirement depends on the purpose to which the solution will be applied.
  Furthermore, convergence is also often measured by the level of residuals, the amount by which discretised equations are not satisfied, and not by the error in the solution. The user should therefore be aware of this, in deciding what convergence criterion should be used to assess a solution.
- *Consistency* is based on the concept that numerical schemes must produce systems of algebraic equations which can be equivalent to the original governing equations as the grid spacing tends to zero.
  Thus, a discretisation scheme is called consistent, if the discretised equations converge to the given differential equations for both the time step and grid size tending to zero. A consistent scheme ensures that only governing equations are solved. Furthermore, the process to evaluate the appropriateness of the solving equations is called verification which should not be confused with the term validation. Verification tries to quantify the numerical errors, whereas validation deals with the modelling errors. The consistency of a numerical scheme can be checked by expanding the function values into Taylor series. The developments are then inserted back into the discretised equations. If the differential equations are subtracted, terms which represent the numerical error are obtained—the so-called truncation error. For a consistent scheme, the truncation error should go to zero with decreasing time step and grid size (Blazek 2008).
- *Stability* is linked to damping error as the numerical method proceeds. If a technique is not stable, even round off errors in the initial data can cause wild oscillations or divergence.
  The conditions under which a given numerical scheme converges can be known a priori by performing a stability analysis of the scheme. A numerical method is referred to as being stable when the iterative process converges and as being unstable when it diverges. It is not possible to carry out an exact stability analysis for Navier-Stokes equations. But a stability analysis of simpler, model equations provides useful insight and approximate conditions for stability.
  A common strategy used in CFD codes for steady problems is to solve the unsteady equations and to march in time until the solution converges to a steady state. A stability analysis is usually performed in the context of time-marching. While using time-marching to a steady state, the interest is focused on accurately obtaining the asymptotic behaviour at large times. So the procedure consists of taking as large a time-step (Δt) as possible to reach the steady state in the least

number of time-steps. There is usually a maximum allowable time-step (Δt_{max}) beyond which the numerical scheme is unstable. If $\Delta t > \Delta t_{max}$, the numerical errors will grow exponentially in time causing the solution to diverge from the steady-state result. The value of Δt_{max} depends on the numerical discretisation scheme used. There are two classes of numerical schemes, explicit and implicit, with very different stability characteristics.

It is important to take into account that convergence is usually very difficult to establish theoretically and, in CFD methods with non-linear governing equations, consistency and stability are necessary conditions for convergence, but not sufficient. Robust finite volume calculation schemes must have three properties: conservativeness, boundedness and transportiveness (Versteeg and Malalasekera 1995).

As mentioned, the finite volume approach guarantees local conservation of a fluid property for each control volume. Numerical schemes with *conservativeness* property also ensure global conservation of the fluid property for the entire domain. It is achieved by consistent expressions for the fluxes of fluid property through the cell faces of adjacent control volumes.

The *boundedness* property is similar to stability and requires that in a linear problem without sources, the solution is bounded by the maximum and minimum boundary values of the flow variable. Boundedness can be achieved by placing restrictions on the magnitude and sign of the coefficients of the algebraic equations. Although flow problems are non-linear, it is important to study the boundedness of a finite volume scheme for closely related, but linear, problems.

Flow processes involve effects due to convection and diffusion. Finite volume schemes with the *transportiveness* property must account for the directionality of influencing in terms of the relative strength of diffusion to convection.

Thus, conservativeness, boundedness and transportiveness are designed into all finite volume schemes and have been widely shown to lead to successful CFD simulations. Therefore, they have been accepted as alternatives for the concepts of convergence, consistency and stability and it is observed that appropriate CFD frequently involves a balance between solution accuracy and stability.

3.7 CFD Application in Green Design

The increasing demand for environmental engineering and green design has led to the application of CFD in areas such as green building design, wind power, and concentrating solar thermal technologies, among others.

3.7.1 Green Building Design

In the last two decades, CFD has become an useful and attractive tool in building designs and environmental studies. The information provided by CFD can be used to analyse the impact of building exhausts to the environment, to predict smoke and fire risks in buildings, to quantify indoor environment quality, and to design natural ventilation systems. Review of historical literature reveals that, although CFD was introduced to building applications in 1970s, CFD was still new to most building designers and engineers before 1990s (Al-Baghdadi 2014).

Buildings and systems modelled by CFD have become more and more sophisticated, while less knowledge of fluid mechanics and building science is required to conduct CFD simulations. This fact results in the necessity of developing guidelines and standards to regulate the use of CFD for building design in the period 2000–2010. Nowadays, CFD has been used widely in various building design projects and stages, thanks to the enhancement of commercial tool capabilities and easiness as well as the enforcement of some urban and building code regulations.

3.7.2 *Wind Power*

Among renewable sources of energies, wind power is an important source of environmental-friendly energy and has become more and more important in recent years. The number of installed wind power plants is increasing every year and many nations have proposed plans to make large investments in wind power in the near future. In addition, remarkable advances in green design have been possible due to developments in modern technology.

Thus, CFD modelling has been conducted to study the engineering related to wind power such as the flow characteristics around a wind rotor (Fig. 3.1).

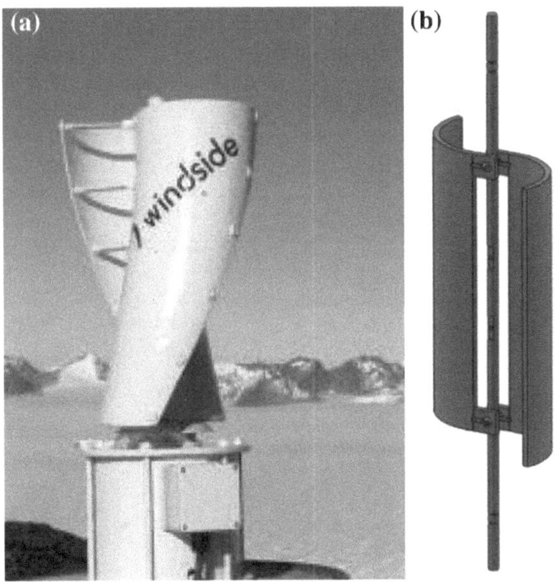

Fig. 3.1 Savonius wind rotor: **a** real geometry, **b** CFD geometry

Numerical simulation leads to the solution of complex flow problems found in wind rotors, because the study of turbulent flow is one of the main challenges in fluid flow. In this type of cases, CFD simulation makes possible to obtain velocity and pressure distributions, together with turbulent characteristics, in different transverse and longitudinal planes of the defined control volume. The information obtained could assist future designs of packaged wind-rotor facilities.

3.7.3 Concentrating Solar Thermal Technologies

The experimental development of concentrating solar facilities requires funding and time to build them. In order to optimise resources and to avoid the assembly of several different facilities, CFD is used as a method to reproduce the operating conditions and analyse the parameters which determine the evaluation and feasibility of the installation.

As mentioned, CFD is one of the most appropriate methods to analyse different phenomena involving fluid flows. In the thermosolar field, CFD modelling is applicable to a wide range of studies. Thus, CFD simulation can be used in the design of new concepts, optimisation of existing designs, thermal analysis of working fluids, and in the optimisation of the operating conditions. In the following chapters, the application of CFD modelling in CST technologies will be studied in detail.

References

Al-Baghdadi MARS (2014) Computational fluid dynamics applications in green design. International Energy and Environment Foundation, Iraq

Bachelor GK (1967) An introduction to fluid dynamics. Cambridge University Press, Cambridge

Blazek J (2008) Computational fluid dynamics: principles and applications. Elsevier, Oxford

Patankar SV (1980) Numerical heat transfer and fluid flow. McGraw-Hill, New York

Roldán MI (2012) Design and thermal analysis of a volumetric receiver system installed in a solar furnace. Dissertation, University of Almería

Ueberhuber CW (1997) Numerical computation: methods, software and analysis. Springer, Heidelberg

US Environmental Protection Agency (2015) https://www.epa.gov/green-engineering

Versteeg HK, Malalasekera W (1995) An introduction to computational fluid dynamics: The finite, vol method. Longman Scientific & Technical, Harlow

Chapter 4
Optimisation of Solar Receivers

In solar thermal electricity plants, the receiver is an essential component in which the concentrated solar energy is absorbed and then transferred to a working fluid. Thus, the receiver effectiveness has a significant influence on the overall efficiency of the facility. This chapter deals with the optimisation of solar receivers by CFD simulation, considering both previous designs and operating conditions in the facility. Therefore, several cases are analysed in order to offer a detailed study of the methodology followed.

4.1 Introduction

The improvement of concentrating solar thermal technologies is based on the increase of efficiency in converting the energy from the sun into electricity, while reducing investment costs. The overall efficiency of the plant can be improved by reaching higher working temperatures, but higher temperatures increase the thermal losses of the receiver through convection and radiation, and require materials that withstand these conditions (usually more expensive). However, storage costs can be drastically reduced at higher temperatures, and improved efficiency also lowers the cooling load and the performance penalty caused by dry cooling. Thus, the benefits of working at higher temperatures depend on the concentration ratio.

Linear concentration systems (parabolic troughs and linear Fresnel reflectors) have representative concentration ratios of 40–100, whereas point-focus systems (towers and dishes) have ratios of 100–2000. The efficiency of the receivers depends on the technology, while the maximum possible efficiency of the conversion of heat into electricity is given by Carnot theorem. The ideal Carnot efficiency is defined by the ratio of the difference in temperatures of the hot and the cold sources, divided by the absolute temperature of the hot source.

Hence, the overall efficiency is the product of the receiver efficiency by the Carnot efficiency and a fixed coefficient which defines the imperfection of the

M.I. Roldán Serrano, *Concentrating Solar Thermal Technologies*,
Green Energy and Technology, DOI 10.1007/978-3-319-45883-0_4

thermodynamic engine. It is obtained that point-focus systems can convert into electricity a larger fraction of the energy absorbed by the receiver than linear ones. For each concentration level, there is an optimal temperature which maximises the global efficiency (400 °C for concentration ratios of 100 and 750 °C for concentration ratios of 1000) (IEA 2014). In this context, the optimisation of both the solar receiver design and its operating conditions is essential in order to improve the efficiency of the facility.

4.2 Solar Receivers

4.2.1 Solar Receivers in Medium-Concentration Technologies

In parabolic trough collectors, the receiver consists of a metal tube with a special coating to maximise energy absorption and minimise infrared re-irradiation. In order to avoid convection heat losses at temperatures of around 400 °C, the tubes work in an evacuated glass envelope that is connected with the metal joint by a glass-to-metal seal (Fig. 4.1).

Linear Fresnel reflectors also have a coated tubular receiver, but in this case the evacuated glass envelope is not used because this receiver reaches lower temperatures (270 °C). In order to increase the optical efficiency, the facility includes a secondary reflector that captures the solar beams deviated from the linear focus.

4.2.2 Solar Receivers in High-Concentration Technologies

Solar towers (so-called central receivers) theoretically offer a more efficient design than linear systems, but the achievement of higher temperatures requires greater concentration factors to minimise heat losses in the receiver. In fact, the real efficiency of the receivers varies. Linear systems have greater optical losses in winter

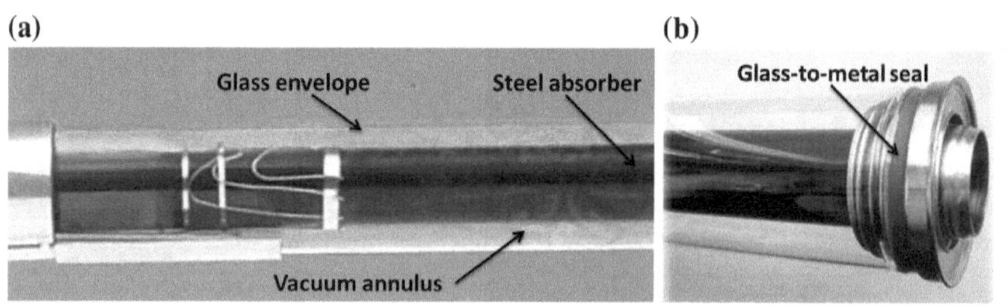

Fig. 4.1 Receiver tube in parabolic troughs: **a** receiver elements, **b** tube connection

and the receiver is insulated by an evacuated glass envelope. However, solar towers are less sensitive to seasonal variations than linear systems and the central receiver does not require a highly effective insulation (IEA 2014).

High-temperature receivers present several challenges, such as the development of new geometric designs to increase the receiver efficiency, materials appropriated to withstand high temperatures, heat-transfer fluids for higher efficiency power cycles, and processes that maximise solar irradiance and absorptance, minimise heat loss, and have high reliability at high temperatures over thousands of thermal cycles (Ho and Iverson 2014).

There are two main high-temperature receiver configurations: external and cavity-type receivers (Fig. 4.2). In the cavity design, the solar flux enters the cavity, ideally closed by a window, though this raises significant material challenges. This design is thought to be more efficient, reducing heat losses, but accepts a limited angle of incoming sunlight, so towers surrounded with large heliostat fields need to support several receivers. External receivers allow the capture of the solar radiation coming from any incident angle, but they need a larger area for surround heliostat fields than the one for north-facing fields (cavity design) and a shorter tower than the required by the cavity-type receiver.

From the analysis of these two configurations, it has been obtained that cavity receivers generally expected to have a lower radiation heat loss and higher convective heat loss than that for external receivers (Falcone 1986). In fact, at elevated receiver temperatures of around 650–750 °C, re-radiation effects must be considered in the selection between an open receiver design and an enclosed one.

Furthermore, there are three concepts of high-temperature solar receivers for the absorption of the incoming radiation: volumetric, tubular and falling absorbent media. The *volumetric-receiver* concept consists of a highly porous material which absorbs solar radiation at different depth through its thickness, showing an effective

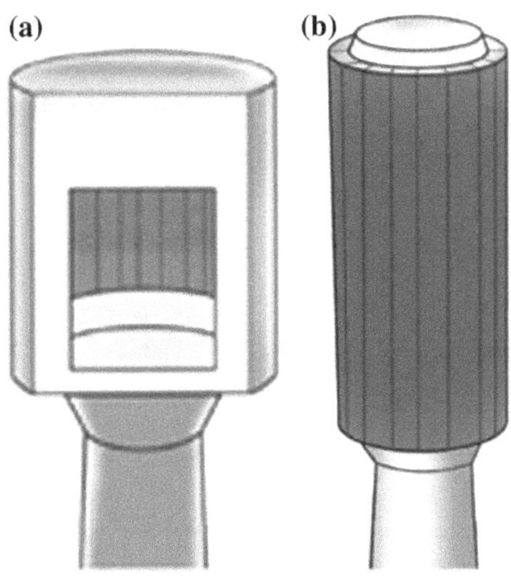

Fig. 4.2 Configurations in high-temperature receivers: **a** cavity, **b** external (Ho and Iverson 2014)

area for solar absorption which is many times larger than that of thermal radiation losses (Fig. 4.3a). Furthermore, a fan sucks a working fluid (usually air) through the absorbent pores, and the convective flow captures the heat absorbed. The outlet fluid temperature should be higher than the temperature of the absorber material on its irradiated surface. These phenomena produce the so-called volumetric effect and results in a minimisation of absorber thermal radiation loss (Hoffschmidt et al. 2003).

Tubular receivers consist of different modules with an array of metallic tubes through which a heat transfer medium flows consecutively. The tubes receive the solar radiation coming from the heliostat field and transfer the heat to the working fluid. Figure 4.3b firstly shows the fluid flowing through the central module and, afterwards, it passes through the lateral ones. It is located a manifold at the inlet and outlet of each module to homogenise the flow conditions (Roldán and Fernández-Reche 2016).

When it is considered a falling absorbent media in a central receiver concept, it can be regarded both a falling fluid film and falling solid particles. *Falling-film receivers* present a gravity-driven fluid motion. The fluid usually flows down an inclined wall, reducing the pumping requirement of the receiver, and it can either be directly irradiated or indirectly heated through the wall. In the case of *falling solid particle receivers*, sand-like ceramic particles fall through a cavity receiver and are directly irradiated by the incoming concentrated solar radiation. The heated particles can be stored in an insulated tank and/or used to heat a secondary working fluid (Ho and Iverson 2014).

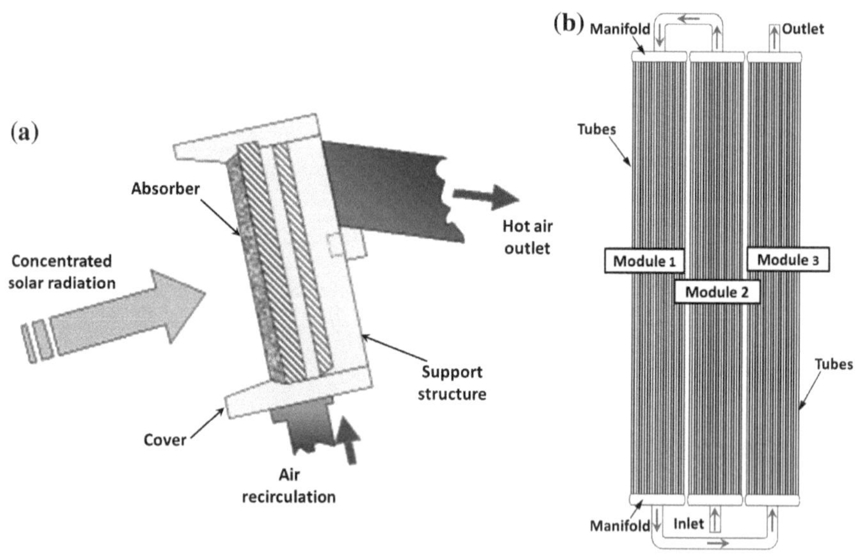

Fig. 4.3 High-temperature receivers: **a** volumetric (Roldán et al. 2015) and **b** tubular receiver (Roldán and Fernández-Reche 2016)

4.3 Optimisation of Solar Receivers

As previously mentioned, concentrating solar thermal technologies require the improvement of their cost competitiveness with respect to other energy conversion technologies by increasing energy conversion efficiencies while maintaining or reducing cost. For that purpose, these technologies should achieve high operating temperatures and heat-flux densities. The receiver is an essential component of the STE plant which, in solar towers, accounts for about the 15 % of the total plant investment cost. With the aim of improving the receiver effectiveness, it is important to study existing receiver designs in order to propose new optimised concepts and it is also necessary to establish the most appropriate operating conditions adapted to each case.

4.3.1 Thermal Analysis and Optimisation of the Receiver Design

As mentioned in Chap. 3, CFD simulation is an useful tool which allows the study and thermal analysis of a fluid-dynamic system. Thus, in this chapter, several examples of volumetric receiver designs are evaluated by CFD simulation in order to obtain a prediction of the receiver thermal behaviour depending on the operating conditions and to propose alternative designs which can optimise the receiver efficiency.

In this case, it has been considered the volumetric central-receiver concept because high-temperature receivers reach the highest efficiencies, the volumetric design is not well developed and the use of atmospheric air as heat transfer fluid involves advantages in terms of availability, cost and environmental impact compared to other commercial fluids.

This type of studies requires the evaluation of certain parameters in order to determine the adequacy of the receiver design to the needs of the facility. The main parameter is the receiver thermal efficiency together with the analysis of temperature, velocity and pressure distribution in the selected domain.

4.3.2 Operating Conditions

Some operating conditions in solar towers cannot always be controlled. The incoming radiation depends mainly on weather conditions and it is important to ensure a proper aiming strategy by the heliostat field's control system in order to optimise the solar flux map on the receiver, thereby allowing the highest solar input while avoiding any local overheating on the receiver area (IEA 2014). Furthermore, it is important to take into account the wind effect on the front surface of the receiver and how it can be attenuated by considering other operating parameters such as the recirculation rate and the temperature of the recirculation air.

These operating conditions can also be studied by CFD simulation which, once validated, leads to establish a prediction procedure that enables the study of the receiver behaviour under any operating conditions. In this chapter, it is shown a detailed analysis of the wind effect and the influence of the recirculation air on a volumetric receiver installed in a solar tower.

4.4 Thermal Analysis and Optimisation of a Solar Volumetric Receiver by CFD Simulation

4.4.1 Thermal Analysis of a Solar Volumetric Receiver

In this section, the thermal analysis of a metallic volumetric receiver by CFD simulation is described. Its operating pressure is the atmospheric one, it consists of several absorber modules supported by a hexagonal structure and the porous material is a metallic mesh on which the incoming solar radiation impinges. In this case, there is a low pressure drop in the absorber due to its high porosity (97.4 %) and it makes possible to reach a homogeneous distribution of air flow using a series of regulating orifices located on the back surface of each absorber module (Haeger et al. 1994).

The complete receiver (Fig. 4.4a) implies a complex geometry. In order to minimise the computational cost, a two-dimensional geometry has been considered by the cross profile at the centre of the receiver aperture, regarding seven absorber modules crossed by their central diagonal. Figure 4.4b shows the selected domain where the subdomains are identified.

The CFD model can be validated by comparison between simulation results and experimental data obtained from the steady state determined for two tests whose conditions are collected in Table 4.1.

The mesh selected is made up of quadrilateral cells because structured grid favours convergence. In order to select the best mesh size, it is recommended to develop a mesh independence test to adjust the grid size to the minimum one that ensures an independent solution of the cell number. Furthermore, the mesh quality has been evaluated by the *equiangle skew* (Q_{EAS}) and the cell deviation from the equilateral shape is obtained by the *aspect ratio* (Q_{AR}). The 100 % of the cells are in the Q_{EAS} range 0–0.4 which corresponds to both the excellent quality range (0–0.25) and the good one (0.25–0.5). The maximum Q_{AR} is 1.28 that is close to 1, thus there is a slight deviation from the equilateral shape. Figure 4.5 shows the adaptation of the mesh selected to the geometry.

The CFD model takes into account the conservation laws described by Eqs. (3.1), (3.2) and (3.3) in Chap. 3. These general equations were regarded in two dimensions and, in this case, the air is the only species involved in the fluid

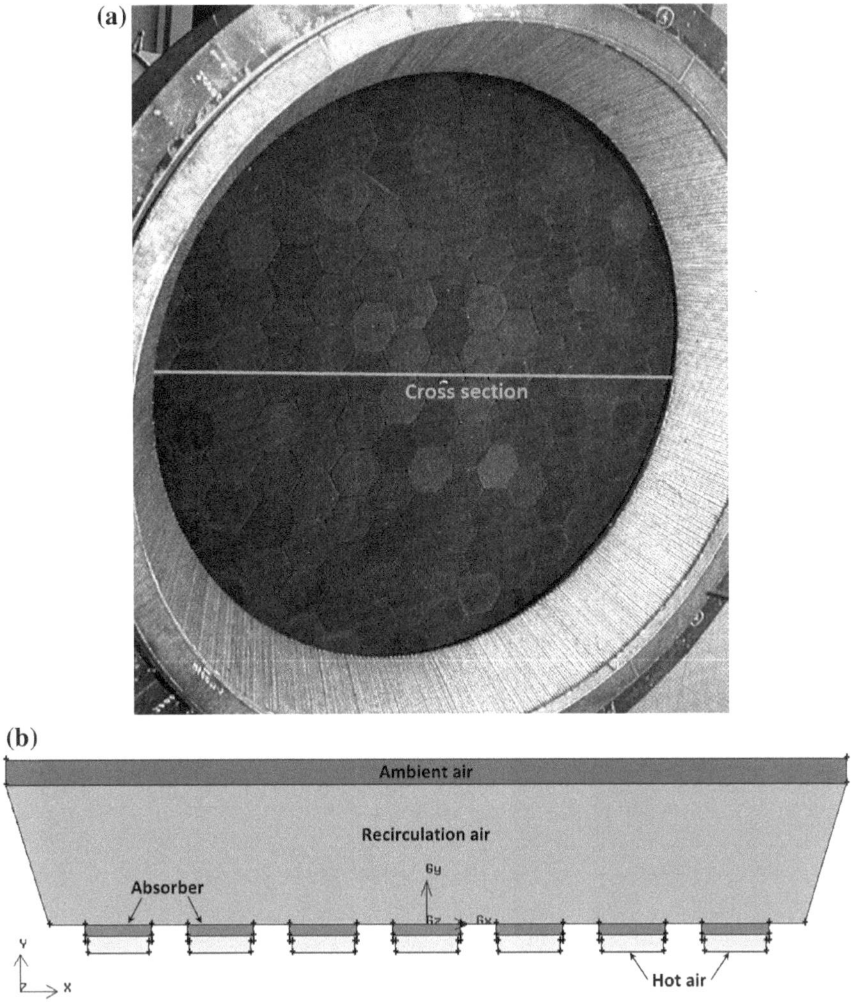

Fig. 4.4 Definition of the 2D geometry: **a** real geometry, **b** selected domain (Roldán et al. 2015)

Table 4.1 Steady-state conditions

Test	Pressure, Pa	Air mass-flow, kg/s	Wind velocity, m/s	Wind direction,°	Total power received, kW
1	97,680	2.72	3.60	142.49	2405.03
2	97,680	3.45	3.30	268.50	2811.54

Source Roldán et al. (2015)

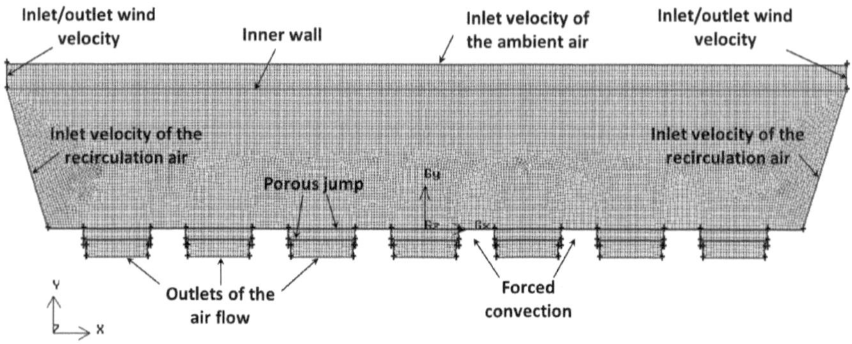

Fig. 4.5 Mesh and boundary conditions for the solution domain (Roldán et al. 2015)

medium. Furthermore, it is assumed a steady-state fluid flow (air), considering the operating conditions summarised in Table 4.1. The porous-medium model takes into account the solid influence on the fluid and it requires to include the solid effect in the energy equation and to consider an additional momentum source term. Therefore, Eq. (3.2) implements for the porous medium a source term composed of two parts: a viscous loss term and an inertial loss term.

$$S_i = -\left(\sum_{j=1}^{2} D_{ij}\mu v_j + \sum_{j=1}^{2} B_{ij}\frac{1}{2}\rho v_{mag} v_j\right) \tag{4.1}$$

where i and j take into account the coordinate directions of the model (x or y), S_i is the source term for the ith (x or y directions) momentum equation (N/m^3), μ is the viscosity of the fluid, ρ is the density of the fluid, v_{mag} is the velocity magnitude (m/s), and D and B are second-rank tensors which are unit tensors for homogeneous porous media.

The energy equation for the porous medium is described by the following expression:

$$\begin{aligned} &\frac{\partial}{\partial t}\left(\gamma\rho_f E_f + (1-\gamma)\rho_s E_s\right) + \nabla\cdot\left(\vec{v}\left(\rho_f E_f + p\right)\right) \\ &= \nabla\cdot\left(k_{eff}\nabla T - \sum_j h_j\cdot\vec{J}_j + \left(\bar{\bar{\tau}}_{eff}\cdot\vec{v}\right)\right) + S_f^h \end{aligned} \tag{4.2}$$

where γ is the porosity of the medium, E_f is the total fluid energy (J/kg), E_s is the total solid energy (J/kg), S_f^h is the fluid enthalpy source term (W/m^3), and k_{eff} (W/m K) is defined by the following equation:

$$k_{eff} = \gamma\cdot k_f + (1-\gamma)\cdot k_s \tag{4.3}$$

where k_f is the fluid-phase thermal conductivity (W/m K), including the turbulent contribution, and k_s is the solid-medium thermal conductivity (W/m K).

The viscous model selected was *κ-ε renormalization group* (RNG) because the Reynolds number evaluated at the absorber outlet for the temperature range of 900–1000 K and considering an experimental mass flow of 0.025 kg/s is low, and this turbulence model accounts for Reynolds-number effects in this range.

In this case, the gravitational force can be neglected because of the low density of the air, the forced air stream and the horizontal position of the receiver. Nevertheless, forced convection has been regarded for the outer and inner walls of each absorber module, using an average heat transfer coefficient of 165 W/m^2 K for air forced convection (coefficient range between 30 and 300 W/m^2 K, Dantzing and Tucker 2001). Furthermore, the mass source term was not taken into account because it is referred to the mass added to the continuous phase from a dispersed second phase (in this case, there is only one phase) and to any user-defined sources.

The heat coming from the solar radiation has been included in the model as a volumetric heat source (Q_v). It was defined for each test depending on the absorber depth (y) by the expressions collected in Table 4.2. They come from an exponential law that is an approach of the radiation-intensity attenuation in the absorber material (Roldán et al. 2014a, b).

The metal mesh of each absorber module is regarded as a porous material defined by the viscous loss term (3.02×10^7 m^{-2}) and the inertial loss term (25.45 m^{-1}) in the flow direction (0, −1). For the secondary flow direction (1, 0), the resistance to the fluid flow is neglected; thereby it is considered a much greater value of these terms (10^{10} m^{-2} and 1000 m^{-1}, respectively).

Figure 4.5 shows the boundary conditions selected in the solution domain. The inlet velocity of the ambient air was obtained from the mass flow measured in the flow direction (0, −1) at ambient temperature (about 295 K). The inlet velocity of the wind was measured together with its incidence angle at ambient temperature. Both air and wind outlets were defined as outflow. The inlet velocity of the recirculation air has been calculated taking into account the inlet area, its density at the inlet temperature (about 520 K) and the recirculation rate that is evaluated from the air mass flow (Romero et al. 1999).

The walls that delimit the porous medium were associated to a porous-jump condition. This case require to include the material permeability (3.314×10^{-8} m^2) and the pressure-jump coefficient (508.82 m^{-1}). The outer walls of the absorber module are defined considering forced convection with a heat transfer coefficient of 165 W/m^2-K and the temperature range of the hot air. The remaining walls are coupled to the adjacent areas.

Table 4.2 Volumetric heat source

Test	Volumetric heat source, W/m^2
1	$Q_v = 5,718,975 \cdot e^{-21.59 \cdot y}$
2	$Q_v = 6,685,775 \cdot e^{-21.59 \cdot y}$

Source Roldán et al. (2015)

The thermophysical properties of the solid materials (absorber and support structure) were implemented in the model, considering the variation of the specific heat capacity with the temperature. The working fluid is dry air, whose properties were defined depending on the temperature (Roldán et al. 2015).

Once defined the simulation model, it was validated by means of two variables: the air temperature at the outlet of the porous material (experimental measurement) and the thermal efficiency of the absorber (experimental value obtained from real measurements). Therefore, the measurements of the air temperature at the absorber outlet have been compared with the simulation data considering the measurement uncertainty (range between ±1.1 and ±3.1 K) and the simulation uncertainty determined by the flowmeter (uncertainty of ±5 %), because the mass flow is used as boundary condition.

According to Table 4.3 and considering the air temperature measured at the recirculation area (T_r) and at the outlet of the absorber (T_{abs}) for both tests, the maximum deviation is lower than 3.5 %. Thus, it can be concluded that simulation results are in appreciable agreement with experimental results.

Furthermore, the thermal efficiency achieved was evaluated by the following equation (Roldán 2013):

$$\eta = \frac{Q_{conv}}{Q_{rec}} \quad (4.4)$$

where η is the thermal efficiency of the receiver, and Q_{conv} is the convective flow obtained from the energy balance between the fluid inlet and the absorber outlet (W):

$$Q_{conv} = m_f \cdot cp_f \cdot \left(T_{f,out} - T_{f,in}\right) \quad (4.5)$$

where m_f is the air mass flow that passes through the absorber modules (kg/s), cp_f is the average specific heat capacity of the air (J/kg K), $T_{f,in}$ is the air temperature at the absorber inlet (K), and $T_{f,out}$ is the air temperature at the outlet of the absorber module (K).

Q_{rec} is the heat absorbed from the incoming concentrated solar radiation over the inlet receiver surface (W), whose evaluation considers the superficial heat source (I_{SO}, W/m^2) according to the frontal receiving area of 7.07 m^2 (A).

$$Q_{rec} = I_{SO} \cdot A \quad (4.6)$$

Table 4.3 Deviation between experimental data and simulation results

Test	$(T_{abs})_{exp}$, K	$(T_{abs})_{sim}$, K	Deviation, %	$(T_r)_{exp}$, K	$(T_r)_{sim}$, K	Deviation, %
1	981.7	1014.7	3.36	380.1	381.9	0.47
2	997.0	1003.8	0.68	369.8	363.5	1.71

Source Roldán et al. (2015)

Table 4.4 Deviation between experimental data and simulation results

Test	$(T_{f,out})_{sim}$, K	$T_{f,in}$, K	cp_f, J/kg-K	m_f, kg/s	Q_{conv}, kW	I_{so}, W/m^2	Q_{rec}, kW	η_{sim}, %	η_{exp}, %	Deviation, %
1	1014.7	562.2	1047.4	2.7	1289	264,890	1873	68.8	69.5	0.96
2	1003.8	521.0	1038.8	3.4	1730	309,670	2189	79.0	79.6	0.71

Source Roldán et al. (2015)

From the thermal efficiency evaluation, it has been obtained that the maximum deviation between the experimental one and the efficiency evaluated from the simulation results is around 1 %, which can be considered as a good approach (Table 4.4).

As a summary of this section, this study describes the methodology followed to develop a simplified CFD model able to analyse the thermal behaviour of a metallic volumetric receiver. The reliability of the model is demonstrated by comparing the experimental air temperature, measured at both the outlet of some absorber modules and the recirculation area, with the simulation data obtained at the same location for the tests selected. The maximum deviation obtained was lower than 3.5 %, which can be considered as a good approach.

Furthermore, the thermal efficiency obtained from the simulation results was studied taking into account the overall thermal efficiency of the receiver. Its comparison with the experimental one resulted in a low deviation (around 1 %), obtaining a thermal efficiency range from 69.5 to 79.6 %. As a consequence, the CFD model developed is reliable and able to predict the thermal behaviour of the receiver under any operating conditions selected by the user, providing a useful tool for the thermal analysis of the receiver design considered.

4.4.2 Design Optimisation for a Solar Volumetric Receiver

As an example of solar receiver optimisation, it has been selected a ceramic volumetric receiver at atmospheric pressure coupled to a process chamber for thermal treatments. The CFD model developed allows the evaluation of the initial receiver configuration with different constant-porosity absorbers and, from the thermal efficiencies obtained, new designs with gradual porosities are proposed and analysed.

These configurations were proposed for the same experimental absorber material (Silicon Carbide) and they were divided into three groups, as it is summarised in Table 4.5. One of them considered three configurations with different constant porosities (0.48, 0.64, and 0.78), the other group took into account two configurations with a gradual porosity variation in the radial direction using the previous three constant porosities in increasing and decreasing order, and the last group included two gradual configurations according to the absorber depth (6 cm) using also the increasing and decreasing order. The simulation parameters were evaluated

Table 4.5 Configuration proposed

Constant porosities	Gradual porosities in the radial direction		Gradual porosities according to the absorber depth	
0.48	0.48	0–3.5 cm	0.48	0–2 cm
	0.64	3.5–7 cm	0.64	2–4 cm
	0.78	7–10.5 cm	0.78	4–6 cm
0.64	0.78	0–3.5 cm	0.78	0–2 cm
	0.64	3.5–7 cm	0.64	2–4 cm
0.78	0.48	7–10.5 cm	0.48	4–6 cm

Source Roldán et al. (2014a, b)

by means of the reference values calculated for the configuration which was used in the experimental setup (constant porosity of 0.64).

The numerical model is defined by the same equations described in Sect. 4.4.1. It is also considered a two-dimensional CFD model, the air is the only species involved in the fluid medium and it is assumed a steady-state flow condition for the fluid. Therefore, an experimental quasi-steady state obtained from each absorber test, considering its initial design, was defined in order to calculate the steady mass-flow forced by a blower (0.013 kg/s).

Laminar and transitional flow regimes were present. The flow regime was considered laminar in the absorber domain (porous medium), but the RNG κ-ε viscous model was used in the fluid domain of the receiver entrance because, as in previous case, accounts for low-Reynolds-number effects. The near-wall region was calculated by the *wall function approach* which is a solution approach included in the CFD software that regards semi-empirical formulae to obtain the viscosity between the wall and the turbulent region. The gravitational force was also neglected because of the low density of the fluid (air), the forced air stream, and the horizontal prototype position.

A two-dimensional and symmetrical solution domain was considered in this case to define the simplest model capable of predicting the hydraulic and thermal behaviour for gradual-porosity receivers, minimising the computational requirements in comparison with a complex and time-consuming model. This fact makes the reproducibility of experimental cases easier, obtaining results in a shorter time. As symmetrical axis, it was selected the central axis of the system (treatment chamber and volumetric receiver).

Figure 4.6 describes the solution domains for the configurations studied. The sub-domains were selected depending on the boundary conditions involved in the simulation. The inlet-air region was regarded to define the ambient conditions and the air flow rate, two refractory-material domains consider the insulation effect on the model, the porous medium includes the heat transfer from the absorber to the air, and the heated air of the receiver supplies the process heat in the chamber. In the constant-porosity configuration, the porous medium is regarded as a single sub-domain. However, in the case of the gradual porosity, the absorber domain was divided into different sub-domains depending on the porosity (Fig. 4.6b, c).

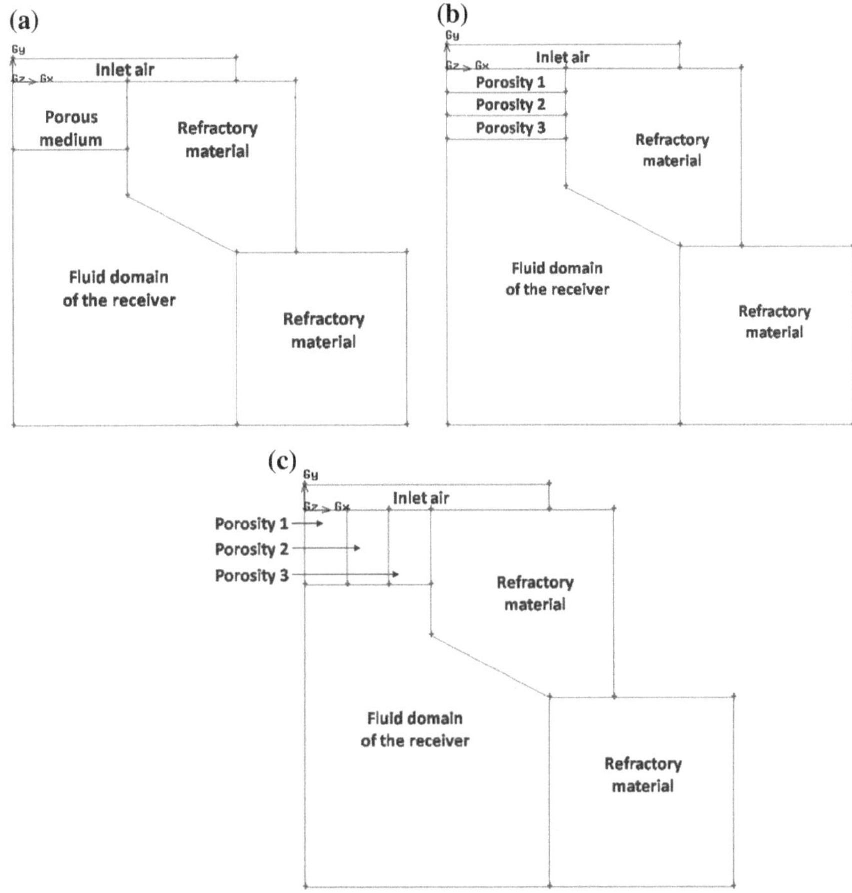

Fig. 4.6 Description of the solution domain: **a** constant porosity, **b** gradual porosity with the depth, **c** gradual porosity in radial direction (Roldán et al. 2014a, b)

The solution domain was discretised by a 2D structured mesh of quadrilateral elements. Similarly to the previous case, the mesh quality was analysed, obtaining that 100 % of the cells have a value of Q_{EAS} included between the values of 0 and 0.5. As mentioned, this range represents a good mesh quality and, in general, it is considered that 2D high-quality meshes should contain elements that possess average Q_{EAS} values of 0.1. In this case, the average Q_{EAS} is 0.13, and it is quite close to the value which determines the high-quality mesh. Thus, this mesh analysis led to select the designed grid for the simulations.

The thermophysical properties of the air stream were defined for an average relative humidity of 43 % at the temperature of 296 K. These averaged conditions were obtained from the measurement of ambient air during the test. The air properties were described by equations derived from the available moist air studies.

The properties of the solid materials were obtained from their data sheets and were implemented in the software library.

The continuum zones defined in the solution domain were porous-medium, refractory material of the absorber module, inlet air, refractory material of the receiver, and the fluid of the receiver. In case of gradual porosity, three different sub-domains were considered in the porous medium, one per porosity (Fig. 4.6).

In the porous-medium continuum zone, it is implemented the volumetric heat source defined by an user defined function (UDF), the viscous loss term and the inertial loss term. These two parameters were evaluated experimentally by the measurement of the pressure loss depending on the air velocity when it passes through the absorber-material depth. These variables were related one to another by the Darcy's law in the Forchheimer extension for one direction, obtaining the viscous loss term (1.6×10^7 m^{-2}) and the inertial loss term (52.63 m^{-1}). These parameters were considered in the flow direction (0, −1) and a much greater value was regarded (10^{10} m^{-2} and 1000 m^{-1}, respectively) in the direction (1, 0) to enhance the fluid flow in the main direction (0, −1) (Roldán et al. 2014a, b).

The volumetric-heat-source definition was based on the exponential law mentioned in Sect. 4.4.1. This phenomenon was described by the following equation (Becker et al. 2006):

$$I(x,y) = I_0 \cdot \mathrm{e}^{-\xi \cdot y} \tag{4.7}$$

where I is the intensity of the solar radiation which goes through the absorber depth (W/m^3), ξ is the optical extinction coefficient that represents the attenuation of the incoming radiation when passes through the porous material (m^{-1}), x is the position in x-axis direction (m), y is the position in y-axis direction (m), and I_0 is the initial intensity of the solar radiation (W/m^3). I_0 depends on the incoming superficial heat flux, which is evaluated from the optical properties of the solar furnace facility where the receiver is installed.

The volumetric heat source for the reference configuration (constant porosity of 0.64) was defined by the following equation considering the Gaussian distribution of the concentrated solar radiation on the absorber surface:

$$I(x,y) = \left(953{,}500 \cdot \exp\left(\frac{-1}{2} \cdot \left(\frac{x^2}{0.0724^2}\right)\right)\right) \cdot 0.849 \cdot 12.35 \cdot \mathrm{e}^{-12.35 \cdot y} \tag{4.8}$$

where ξ is 12.35 m^{-1}. The extinction coefficients for the new porosities considered (0.48, 0.78) were calculated from the comparison between the heat-transfer coefficient of the reference porosity and those coefficients obtained for 0.48 and 0.78. These heat-transfer coefficients were estimated from the simulations under the same conditions as the reference-porosity simulation. The extinction coefficient obtained for the porosity of 0.48 was 11.02 and 13.15 m^{-1} for the one of 0.78.

In radial direction, the porous domain consisted of three sub-domains located as concentric rings (Fig. 4.6c). Consequently, the Gaussian function cannot be regarded in the definition of I_0 because three maximum temperatures would be interpreted, one for each sub-domain, due to the independent definition of the volumetric energy source for each porous medium. To avoid that, an average value (456,000 W/m^2) substitutes the Gaussian superficial heat flux in the equation of the volumetric heat source:

$$I(y) = 456{,}000 \cdot \xi \cdot e^{-\xi \cdot y} \tag{4.9}$$

where ξ is replaced by the value obtained for each porosity (11.02, 12.35 and 13.15 m^{-1}, for porosities of 0.48, 0.64, and 0.78, respectively). Thus, three UDFs had to be implemented in the model. For that purpose, a UDF library was created in order to simultaneously include the three equations. Therefore, the gradual-porosity configuration according to depth used the same UDF library for making a comparison between both gradual configurations.

The porous-jump boundary condition was defined for those walls which delimit the porous medium. This case required the definition of both the material permeability (6.11 × 10^{08} m^2) and the pressure-jump coefficient evaluated from the second coefficient of the fitting function for the Forchheimer's law (104.73 m^{-1}).

The velocity inlet condition was used in the inlet air domain with a value of 0.332 m/s. This velocity comes from the evaluation of the mass flow at ambient temperature in the steady state selected. The direction of the flow was defined with the vector components (0, −1) in the Cartesian coordinate system.

Furthermore, the walls connected with the insulating material, which has not been regarded in this simulation, have been defined as adiabatic walls. A convective condition was defined in the outer wall, and the average heat-transfer coefficient because of natural convection was fixed (13.25 W/m^2 K). Finally, the inner wall of each domain was coupled with the remaining boundary zones.

With regard to the solver method, it was selected the segregated steady-state one, and variable residuals were monitored to ensure the convergence criterion. They were fixed at 10^{-6} for the energy, and at 10^{-3} as a minimum for the continuity and momentum variables. The pressure discretisation scheme used was *standard*, and the algorithm for the pressure-velocity coupling chosen was *simple*, which is the most commonly used in cases without complex phenomena such as high levels of turbulence.

The validation of the CFD model considered experimental measurements of the air temperature at the porous-material outlet and simulation results at the same location, taking into account the steady state of a reference-absorber test. The deviation between simulation results and experimental values was in the range of 1.7–2.8 % (16.7–27.44 K). Therefore, the CFD model developed can predict the absorber behaviour with a maximum deviation of 3 % from the experimental values, including the deviation produced by the measuring instrument (measuring accuracy of ±3.9 K at temperature ranges of 964–981 K).

This difference is due to the difficulty of taking high-temperature measurements in a fluid. In addition to the measuring error, both the thermal radiation and the fluid flow influence the temperature values obtained. Furthermore, the deviation can also be produced by the assumptions of the 2D numerical model and the definition of a volumetric heat source which includes the radiation heat transfer and the heat absorbed by the porous structure. For this reason, a dimensionless evaluation was used, obtaining that numerical results are in appreciable agreement with the experimental data for deviations lower than 5 %.

Regarding the thermal analysis of the three constant porosities (0.48, 0.64 and 0.78), simulation results showed that the maximum temperature (705.4 K, Table 4.6) is reached by the highest porosity, because it allows a higher penetration of the incoming radiation into the absorber structure and presents a greater heat-exchange surface. It was also observed that the absorber material achieves a maximum temperature at the centre of the structure.

Thus, in order to minimise the overheating effect, two configurations with gradual porosity in radial direction were analysed. One of them consisted of an increasing porosity from a value of 0.48 at the centre of the porous medium up to an external porosity of 0.78, and the other configuration considered a decreasing porosity.

These configurations present a maximum temperature concentrated in a smaller area at the centre of the absorber structure (Fig. 4.7c, d), and it corresponds to 938 K for increasing porosity and 1323 K in the case of decreasing porosity in radial direction (Table 4.6). The analysis of the temperature and velocity profiles shows that the decreasing configuration presents the greatest gradient for both variables, being the velocity gradient lower than the thermal one. The temperature difference was around 58 % considering the maximum and minimum values obtained from the simulation results at the absorber outlet of the decreasing configuration, and it shows the greatest overheating at the absorber centre in comparison with constant-porosity configurations. Therefore, the increasing-porosity

Table 4.6 Comparison between the gradual-porosity configurations considering an average value for the superficial heat source

Configuration	Porosity	Maximum temperature, K	Thermal gradient at the absorber outlet, K	Minimum velocity at the absorber centre, W	Velocity gradient at the absorber outlet, K
Constant porosity	0.48	634.7	60.7	1.139	0.63
	0.64	678.1	79.0	1.210	0.67
	0.78	705.4	96.3	1.241	0.72
Gradual porosity in radial direction	Increasing	937.6	315.6	1.344	0.62
	Decreasing	1323.1	758.6	1.349	0.71
Gradual porosity according to depth	Increasing	785.4	108.4	1.521	0.56
	Decreasing	829.3	150.7	1.537	0.65

Source Roldán et al. (2014a, b)

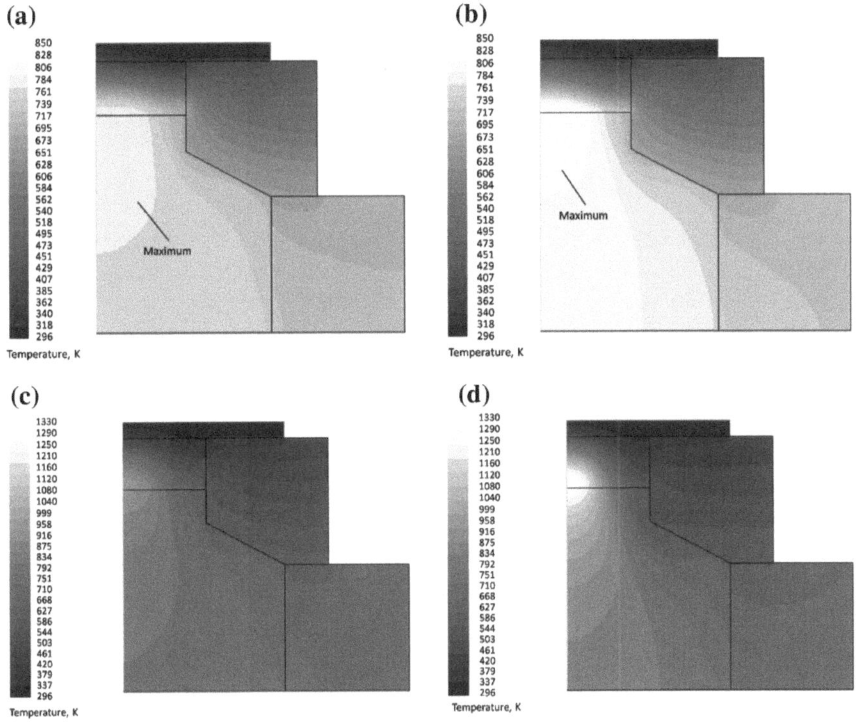

Fig. 4.7 Thermal profile for alternative configurations: **a** increasing porosity with the depth, **b** decreasing porosity with the depth, **c** increasing porosity in radial direction, **d** decreasing porosity in radial direction (Roldán et al. 2014a, b)

configuration presented a more favourable behaviour due to the inherent limitations of the porous material. In this case, the increasing porosity had a lower thermal gradient than the decreasing one, thus it allows the preservation of the absorber material from the thermal shock.

Additionally, two different configurations were regarded to analyse the effect of a gradual porosity according to depth. One considered an increasing porosity (0.48–0.64–0.78) and, the other, a decreasing one (0.78–0.64–0.48). In order to compare the gradual porosity in the radial direction with the one according to depth, the same average value for the superficial heat source was considered (Eq. 4.9). Simulation result (Table 4.6) showed that the maximum temperature is reached by the decreasing porosity (829 K) located in a smaller area at the centre of the cavity compared to the increasing one (Fig. 4.7a, b), which presents a lower temperature (maximum temperature of 785 K).

The difference between the two average-temperature values at the absorber outlet is around 4 % (Table 4.7) and the velocity variation is 3.5 % (Table 4.6), with a greater gradient for both variables in the decreasing configuration. In both cases, the velocity at the centre presents a minimum, because, in the warmer area, the

Table 4.7 Evaluation of the thermal efficiency considering an average value for the superficial heat source

Configuration	Porosity	T_{in}, K	T_{out}, K	Q_{conv}, W	Q_{rec}, W	η, %
Constant porosity	0.48	378.9	621.6	21,083	43,959	47.96
	0.64	394.9	660.3	23,099	44,565	51.83
	0.78	406.4	683.1	24,118	44,840	53.79
Gradual porosity in radial direction	Increasing	434.8	791.5	28,702	43,183	66.47
	Decreasing	502.1	908.7	33,139	46,225	71.69
Gradual porosity according to depth	Increasing	384.4	756.2	32,730	44,287	73.90
	Decreasing	375.7	788.1	36,274	44,243	81.99

Source Roldán et al. (2014a, b)

permeability is lower and the air viscosity is higher. Both effects avoid the absorber cooling. This fact shows that this area is overheated, but this central velocity is greater than the one obtained for the radial configuration (1.5 and 1.3 m/s, respectively). The results reveal that gradual porosity according to depth reaches a lower maximum temperature. However, the thermal distribution for this case is more homogenous than the one for the gradual porosity in radial direction, because the radial variation avoids a homogeneous heat-flux distribution. It enhances the Gaussian effect of the irradiance peak at experimental conditions, strongly limiting the absorber durability in the overheated region.

For the purpose of obtaining the best design for the solar volumetric receiver, the thermal efficiency was evaluated for each configuration by the equations described in Sect. 4.4.1. From the constant porosity evaluation, it is obtained that he highest porosity achieves a thermal efficiency of 53.79 % (Table 4.7) together with a thermal gradient of 96.3 K and a velocity variation of 0.72 m/s at the absorber outlet (Table 4.6). Thus, the configuration with the porosity of 0.78 increases the efficiency in around 6 % and the maximum temperature is improved in around 71 K compared to the lowest porosity (0.48).

Combinations of the three constant porosities were proposed in order to improve the thermal efficiency obtained. All gradual-porosity configurations presented a higher thermal efficiency in comparison with the constant-porosity ones, and the efficiency difference was between 12 and 28 % with regard to the constant porosity of 0.78.

Decreasing-porosity configurations achieved greater thermal efficiencies than those obtained from the increasing ones (around 5–8 % higher), and gradual porosities according to depth reached the highest efficiencies (Table 4.7) with a lower thermal gradient than the resulting one in the porosity radial variation (Table 4.6).

Therefore, the decreasing porosity according to depth was proposed as the best design, which achieves a more homogeneous heat-flux distribution and the highest thermal efficiency (81.99 %) with thermal and velocity gradients of 150.7 K and 0.65 m/s, respectively. These results showed that the use of a greater porosity in the area which receives the solar radiation allows the reduction of the heat losses to the environment because the solar beams reach a greater penetration depth into the porous material. Thus, as a result of the CFD analysis, the decreasing-porosity configuration according to depth was proposed as an alternative to the constant-porosity design (Roldán et al. 2014a, b).

4.5 Optimisation of the Operating Conditions for a Solar Volumetric Receiver by CFD Simulation

The influence of the operating conditions on central receivers, such as wind velocity and return-air conditions, has not been analysed in detail, apart from the study of different air return modes developed by Marcos et al. for an open volumetric receiver (Marcos et al. 2004). This is an important issue to solve for the suitable operation of a solar tower. In this section, CFD simulation is used to analyse the influence of the wind and the effect of the return-air conditions on the efficiency of a high-temperature receiver. This system is a ceramic open solar volumetric air receiver, which consists of a modular ceramic absorber, a supporting structure and an air-return system (Hoffschmidt et al. 2003). The study of these variables enables the definition of the proper strategy to operate the solar-tower facility, improving the thermal efficiency of the receiver.

For that purpose, a CFD model has been developed and simulation results must be compared with experimental data in order to evaluate the model reliability. As mentioned, volumetric receivers of solar towers consist of several absorber cups. This configuration implies a complex geometry that can be simplified by considering a 2D geometry which also minimises the computational cost.

4.5.1 Facility Description

The solar-tower receiver studied is a volumetric air system for high temperature and consists of an absorbent part with 32 porous honeycomb modules located onto its ceramic cup-shaped supports, fixed by a stainless-steel structure, and an air-return system to keep the metallic structure cold in order to avoid the thermal stress (Fig. 4.8). The receiver is 0.9 m in external diameter and presents a rectangular aperture composed of the 32 hexagonal absorber modules, which make it possible to obtain a thermal power of 250 kW (Hoffschmidt et al. 2003).

The absorber material is rccrystallized silicon carbide (SiC) with a total absorber area of 0.41 m^2 and a porosity of 49.5 %. Figure 4.8 shows that the cooling air (ambient air and return air with an air return ratio of around 30 %) passes from the back side of each cup to the front part of the absorber, through the separation channel between the cups. This air stream is mixed with the ambient air in the area close to the absorber aperture and it is partially reabsorbed by the absorbing matrix,

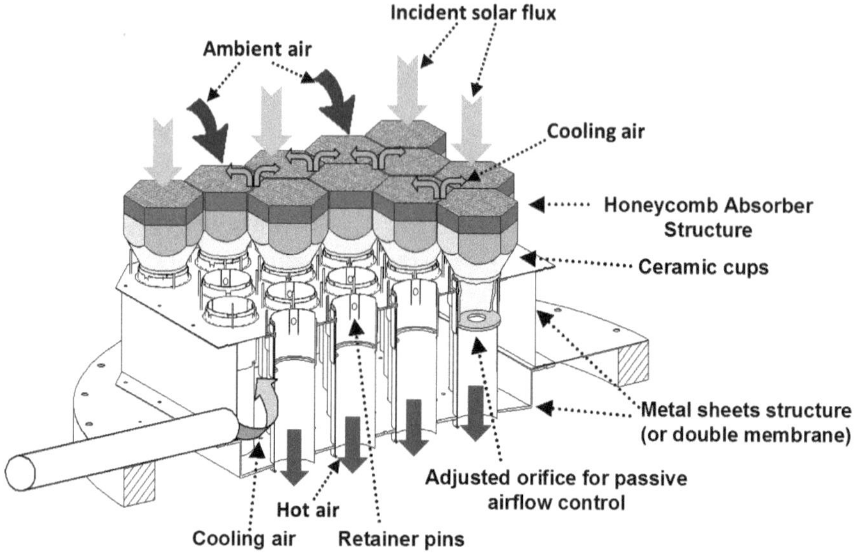

Fig. 4.8 Setup sketch (Roldán et al. 2016)

where the fluid is heated by convection. Furthermore, the receiver design presents a passive control of the airflow by orifices adjusted to the air mass-flow rate. The air flow obtained allows the achievement of homogeneous outlet air temperatures from the cups (Hoffschmidt et al. 2003).

4.5.2 *Procedure*

The prediction of the thermal behaviour of a solar volumetric receiver leads to the optimisation of its operating conditions and, as a consequence, the improvement of its efficiency. Thus, the objective of the 2D CFD model is the selection of the most appropriate operation strategy. The wind parameters (magnitude and direction) and the return-air conditions (temperature and recirculation rate) have been considered in order to determine their influence on the outlet air temperature.

First of all, the model must be validated using experimental data obtained from the test facility. A constraint of central receiver systems is that they present quasi-steady states instead of steady states due to both the variability of the direct normal radiation and the inherent limitations of the control system. Hence, in the validation process, the quasi-steady states for five tests (Table 4.8) were selected according to the definition considered by Hoffschmidt et al. The response measurement used to check the stationary conditions were the air temperature at the receiver outlet and the power gained in the air flow at the same location which depends on both the air temperature and the air flow.

Table 4.8 Test conditions for the cases considered in the validation process at quasi-steady state

Test	Local time, h:min:s	Pressure, Pa	Air flow, kg/s	Wind velocity, m/s	Wind direction,°	Total incident power, kW
1	13:16:07	96,300	0.264	4.20	25.06	263.4 ± 14.2
2	11:31:48	96,400	0.246	1.50	22.73	233.3 ± 12.6
3	12:33:01	96,300	0.231	2.55	21.28	247.9 ± 13.4
4	11:38:29	96,100	0.198	3.70	22.71	253.1 ± 13.7
5	11:06:39	95,400	0.310	5.15	23.86	233.2 ± 12.6

Source Roldán et al. (2016)

In order to determine the influence of the wind and the return air conditions on the outlet air temperature, several values of each parameter have been considered at different ranges: 0–7 m/s for the wind velocity, 0–90° for wind direction (where the value 0° corresponds to the perpendicular direction to the receiver surface), 300–600 K for the return-air temperature, and 0.64–1.34 m/s for the return-air velocity. Table 4.9 includes the values selected for the evaluation of both the wind and the return air.

The heat flux received on the absorber aperture has been measured to consider the irradiance distribution in the volumetric heat-source definition for each absorber cup. The 2D volumetric-heat-source model has been obtained from the fitting of the values evaluated for the total intensity of the solar radiation, considering the maximum incidence angle of the radiation coming from the heliostat field.

As expected, the 2D model is correlated with an exponential model (Eq. 4.7, Roldán et al. 2014a, b), showing a radiation attenuation according to the Beer-Lambert law. Re-radiation effects on the first part of the channel are taken into

Table 4.9 Values selected for the evaluation of the wind and the return air

Wind velocity, m/s	Wind direction,°	Return-air temperature, K	Return-air velocity, m/s
0	0	300	1.34
3.5	45	400	0.64
7	90	500	
		600	

Source Roldán et al. (2016)

Table 4.10 Definition of the volumetric heat source (I_v)

Test	Volumetric heat source, W/m^3
1	$I_v = 1304 \cdot e^{-295.4 \cdot y}$
2	$I_v = 1155 \cdot e^{-295.4 \cdot y}$
3	$I_v = 1227 \cdot e^{-295.4 \cdot y}$
4	$I_v = 1253 \cdot e^{-295.4 \cdot y}$
5	$I_v = 1154 \cdot e^{-295.4 \cdot y}$

Source Roldán et al. (2016)

account by the fitting function which is implemented in the CFD model as a UDF (user defined function) for the volumetric heat source in each absorber structure. The UDFs regarded in the tests are summarised in Table 4.10 and, in all cases, the extinction coefficient is 295.4 m^{-1} which depends on the maximum incidence angle of the radiation and on the pore geometry.

4.5.3 Numerical Modelling

The physics of the model is defined by the three conservation laws (mass, momentum and energy), described in Chap. 3, to analyse the dynamic behaviour of the fluid. This means that the continuity, momentum and energy equations must be solved by the CFD model and they are adapted to both the fluid medium (air) and the 2D geometry considered. In this case, the absorber structure is also considered as a porous medium whose model takes into account the solid influence on the fluid, according to the equations described in Sect. 4.4.1 (Eqs. 4.1, 4.2 and 4.3).

The momentum source term (Eq. 4.1) contributes to the pressure gradient in the porous cell (external body forces); creating a pressure drop that is proportional to the fluid velocity in the cell:

$$S_i = -\left(\frac{\mu}{K_1} v_i + B_2 \frac{1}{2} \rho v_{mag} v_i\right) \tag{4.10}$$

where v_{mag} is the velocity magnitude, ρ is the density of the fluid, μ is the viscosity of the fluid, K_1 is the permeability coefficient and B_2 is the inertial resistance factor, simply specify D and B as diagonal matrices with $1/K_1$ and B_2, respectively, on the diagonals (and zero for the other elements).

The porous media model includes the superficial velocity whose evaluation is based on the volumetric flow rate in the porous region. Thus, the momentum equation for the fluid phase in porous media, considering steady-state, constant porosity, and negligible buoyancy effect is:

$$\nabla \cdot \left(\gamma \rho_f v v\right) = -\nabla(\gamma p) + \nabla \cdot \left(\gamma \bar{\bar{\tau}}\right) + \gamma S_i \tag{4.11}$$

where γ is the porosity of the medium, p is the static pressure, $\bar{\bar{\tau}}$ is the stress tensor and S_i is the source term for the ith (x or y directions) momentum equation (N/m^3).

Furthermore, the porous-medium model also includes the solid effect in the energy equation (Sect. 4.4.1). In this case, the porous medium and fluid flow are assumed to be in thermal equilibrium, the conduction flux in the porous medium uses the effective conductivity defined by Eq. 4.3 and the transient term includes the thermal inertia of the solid region on the medium. The fluid enthalpy source term (S_f^h) corresponds to the volumetric heat source evaluated.

In order to select an appropriate viscous model, the Reynolds number has been calculated for different conditions of the air at the absorber outlet, obtaining a Reynolds-number range of 2947–395 which corresponds to laminar and transitional flow regimes. The viscous model selected was RNG (renormalization-group) k-ε model because it is more accurate and reliable for a wider type of flows than the standard one. This model was derived from the instantaneous Navier-Stokes equations, using a statistical technique called renormalization group theory which provides an analytically-derived differential formula for effective viscosity that accounts for low-Reynolds-number effects. Thus, the potential swirl effects at the outlet of the porous material can be better analysed with this model.

Furthermore, due to the honeycomb structure of the porous material, the flow regime is laminar in the absorber subdomain. This flow regime is defined only for the porous material region, regardless of the overall viscosity model selected. The near-wall region was calculated by the *wall function* approach, in which semi-empirical formulae are used to obtain the viscosity between the wall and the turbulent region. In this case, the standard wall function can be used because the dimensionless distance from the wall (y*) is 16.3 (y* > 15). This distance is defined by the following equation:

$$y^* = \frac{\rho \cdot C_\mu^{1/4} \cdot k_P^{1/2} \cdot y_P}{\mu} \tag{4.12}$$

where k_p is the turbulence kinetic energy at the wall-adjacent cell (P); y_P, the distance from the centroid of the wall-adjacent cell to the wall (P); μ, the dynamic viscosity of the fluid; ρ, the density of the fluid; C_μ, constant of the viscosity model ($C_\mu = 0.0845$).

4.5.4 Geometry Definition and Mesh Design

The solution domain is described in Fig. 4.9, in which the central cross section of the receiver is shown. The passive air-flow control is considered in the geometry, where the central-cup outlets have a greater diameter than the others. As indicated above, two dimensions have been considered in order to define a simple CFD model able to predict the fluid dynamic and thermal behaviour of the receiver, while minimising the computational requirements in comparison with a complex and time-consuming model. This fact allows the obtention of results in a shorter time.

Several subdomains have been defined in the solution domain depending on the boundary conditions. Thus, it was considered the following regions: ambient air which defines the wind and the airflow conditions, the porous medium of each cup which determines the heat transfer from the absorber to the air, the insulating material of each cup that was considered as two different subdomains due to the 2D geometry selected, the hot air which was defined as the area of each cup between

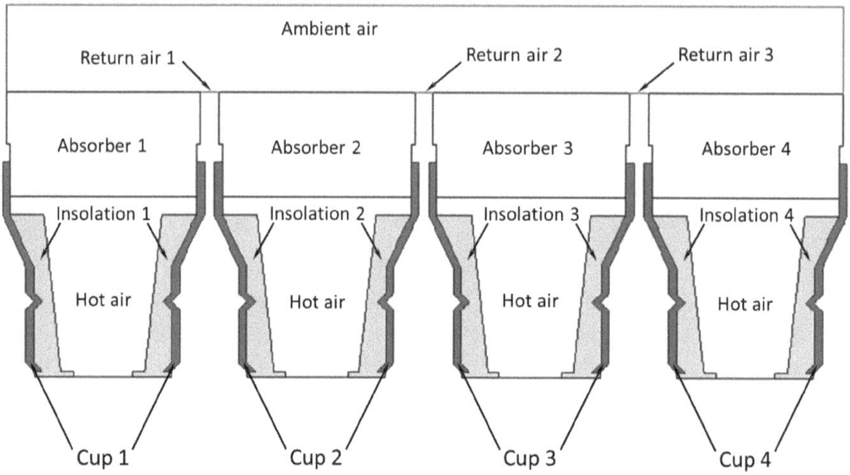

Fig. 4.9 Description of the solution domain (Roldán et al. 2016)

the porous material and the air-flow outlet, and the return air that includes the inlet zone located between the cups. The grid selected for this solution domain consisted of quadrilateral elements (2D structured mesh) and its quality was evaluated by the previously mentioned parameters: equiangle skew (Q_{EAS}) and aspect ratio (Q_{AR}).

The cells included in the good quality range of Q_{EAS} (0–0.5) are 99.98 % of the total quantity and the maximum value obtained is 0.54, close to the range which determines the high-quality mesh. In this case, the maximum value of Q_{AR} is 1.7 and the 87.78 % of the elements are in the Q_{AR} range of 1–1.1, which is very close to the equilateral shape ($Q_{AR} = 1$).

In order to evaluate the appropriateness of the mesh size in areas close to the walls, the y^+ parameter, which determines whether the influences in the wall-adjacent cells are laminar or turbulent, was calculated. In this case, the maximum y^+ was 5.45 and it shows that y^+ values are within the range $0 < y^+ < 6$ corresponding to both viscous sublayer ($y^+ < 5$) and buffer regions ($5 < y^+ < 30$).

Furthermore, the grid resolution has been analysed by a grid independence study adapted to the geometry selected, considering three different mesh sizes. Table 4.11 summarises the results obtained for both the grid resolution and the mesh quality of the three cases. From these data, the mesh selected (23,577 cells) reaches a good

Table 4.11 Grid independency study

Number of cells	$(v_{out})_{Cup1}$, m/s	$(v_{out})_{Cup2}$, m/s	$(v_{out})_{Cup3}$, m/s	$(v_{out})_{Cup4}$, m/s	Minimum orthogonal quality	Maximum orthogonal skew
49,407	5.52	5.21	5.21	5.51	0.651	0.349
23,577	5.74	5.41	5.37	5.67	0.757	0.243
17,108	8.17	7.72	7.71	8.17	0.701	0.299

Source Roldán et al. (2016)

resolution (deviation of the results lower than 3 %), maintaining the best mesh quality (ideal mesh consists of an orthogonal quality of 1 and an orthogonal skew of 0). Therefore, according to Table 4.11, the selected mesh is suitable for the study proposed.

4.5.5 Boundary Conditions

The operating pressure of the solar receiver is the atmospheric one. It is assumed that the fluid (air) is under steady-state flow condition, and the blower of the volumetric receiver produces a flow rate whose value is evaluated for each quasi-steady state condition (0.198–0.310 kg/s). The gravitational force can be neglected due to the low density of the fluid and its forced flow.

This assumption has been studied by the simulation of two cases with a wind velocity of cero and 1.5 m/s, considering buoyancy effects. These results were compared with those evaluated without the gravity effect, obtaining a maximum deviation lower than 1.2 % for the outlet temperature of the cup. Thus, the influence of the gravitational force is not significant in this case.

The thermophysical properties of the air stream have been defined for dry air at atmospheric pressure and the properties of the solid materials were considered at the temperature range of the selected steady states. This information was supplied by the data sheets of each material and the properties were implemented in the commercial CFD code defining new materials in the software database.

As previously mentioned, several continuum zones have been defined in the solution domain: ambient air, the porous medium for each cup, two ceramic areas of each cup, two insulating zones in each cup, and the hot-air zone of each cup (Fig. 4.9). These zones are connected by the definition of different boundary conditions.

Porous-medium continuum zones are described by the UDF of the volumetric heat source, porosity (0.495), the viscous loss term and the inertial loss term (9.06×10^6 m^{-2} and 34.96 m^{-1}, respectively). These parameters have been considered in the direction (0, −1), and, in the direction (1, 0), a much greater value has been regarded (10^{10} m^{-2} and 1000 m^{-1}, respectively) to enhance the fluid flow in the main direction (0, −1) (Roldán et al. 2016).

A porous-jump condition has been selected for the front walls which delimit the porous medium. This condition has been defined with the material permeability and the pressure-jump coefficient (1.1×10^7 m^2 and 139.84 m^{-1}, respectively) evaluated by the Forchheimer's law.

The velocity-inlet condition of the air considers the ambient temperature and the value calculated from the air-flow measured and the absorber front area at the quasi-steady state selected. On the other hand, the direction of the flow was defined with the vector components (0, −1) in the Cartesian coordinate system obtained from the mesh generator.

Table 4.12 Cases considered in the evaluation of the wind influence and the return-air effect

Wind velocity, m/s	Wind direction, °	Return-air velocity, m/s	Return air temperature, K
0	–	1.34, 0.64	300, 400, 500, 600
1.5	45, 90	1.34	300
3.5	0, 20	1.34	300, 400, 500, 600
	45	1.34, 0.64	300, 400, 500, 600
	70, 90	1.34	300
5.5	45, 90	1.34	300
7	0, 20	1.34	300
	45	1.34	300, 400, 500, 600
	70, 90	1.34	300
8	45, 90	1.34	300

Source Roldán et al. (2016)

Wind influence has been taken into account for the lateral side of the ambient-air subdomain. The velocity-inlet condition of the wind (lateral side of the ambient-air subdomain) includes the ambient temperature and the components of the velocity vector according to the Cartesian coordinate system. These values have been evaluated from the velocity and the wind direction measured for the validation process, and, on the other hand, they were set considering the conditions defined in Table 4.12. The other lateral side of the ambient-air subdomain defines the fluid outlet area of the ambient-air subdomain. The return-air velocity inlet is defined by a flow direction of (0, 1), the velocity magnitude is fixed at 1.34 m/s and the temperature is set to 300 K, both for the validation process. The influence of the return-air have been analysed using the conditions collected in Table 4.12.

The outer walls of the cups, absorber structure and insulating material have been defined assuming that the forced convection is the main heat-transfer process. The heat transfer coefficient was set to 30 W/m^2 K (Dantzing and Tucker 2001) and the return-air temperature was included in the wall description.

The forced convection assumption was also used in the definition of the cup inner walls between the absorber outlet and the insulating-material subdomain. In this case, the average temperature of the hot air at the absorber outlet was fixed at 900 K and the value of the heat-transfer coefficient evaluated was 100 W/m^2 K, which is within the range determined for the air circulation by forced convection (30–300 W/m^2 K, Dantzing and Tucker 2001). The lower walls of the insulating material are defined as adiabatic whose heat flux is fixed at zero, and the remaining inner walls were coupled with the appropriate boundary zones.

The developed 2D CFD model solves the governing equations by a segregated steady state solver, and the convergence criterion was fixed at 10^{-6} for the energy, and at 10^{-3} as a minimum for the continuity and momentum variables.

The RNG (renormalization-group) k-ε turbulence-model constants were set to the software's default values and the pressure discretization scheme selected in the software was *standard*. The algorithm of choice for the pressure-velocity coupling was the so-called *simple*. In addition, the convection-diffusion upwinding scheme selected was the *first order upwind* which assumes that the cell-centre values of any field variable represent the cell-average value.

4.5.6 Results and Discussion

The present simulation model has been validated using the temperature measurements obtained from five different tests at the quasi-steady state included in Table 4.8. Each thermocouple has been located at the centre of the cup outlet, whose measurement was compared with the average-weighted air temperature obtained from the simulation at the same location. The results for each test have been independently compared because the operating conditions change in each case.

The measurement uncertainty for the thermocouples used corresponds to an interval between ±2.1 and ±3.5 K in the temperature range measured. The simulation accuracy has been evaluated considering an air-flow uncertainty of ±1 %, and, in order to determine its influence on the simulation results, the outlet-air temperature has been calculated taking into account the maximum and minimum air flow limited by the uncertainty. The temperature variation was around 2.9–4.2 K, thus the air-flow uncertainty is not significant.

The average deviation between experimental and numerical data was lower than 5 % for three cups, but the second one presents a greater deviation, around 26.4 % (Table 4.13). This deviation points out that the temperature at that location was measured mistakenly. In order to verify that assumption, the energy transferred in the solution domain was compared with the heat flux measured, achieving a deviation lower than 5 % (Table 4.13). In this case, CFD model has also revealed a measurement issue in the facility.

On the other hand, the thermal efficiency has been evaluated for each absorber cup using the numerical results by means of the Eq. (4.4), described in Sect. 4.4.1. The numerical thermal efficiency for each absorber cup has been calculated in each test whose values are within a range of 72.1–85.8 %. In order to compare the

Table 4.13 Deviation obtained from the comparison between experimental and numerical data considering the outlet temperature of the fluid and heat flux (Q_{flux})

Test	Dev_{T1}, %	Dev_{T2}, %	Dev_{T3}, %	Dev_{T4}, %	$(Q_{flux})_{sim}$, W	$(Q_{flux})_{exp}$, W	Dev_Q, %
1	9.2	27.2	5.5	5.6	254,515	263,400	3.4
2	2.7	24.2	1.2	1.1	222,745	233,300	4.5
3	0.0	34.6	2.9	0.3	239,490	247,900	3.4
4	7.1	20.9	2.6	3.6	247,358	253,100	2.3
5	5.0	25.1	2.8	3.0	244,560	233,200	4.9

Source Roldán et al. (2016)

Table 4.14 Deviation between experimental thermal efficiency and the one obtained from the simulation results (average value obtained from the four cups)

Test	$\eta_{av\text{-}cups}$, %	η_{exp}, %	Dev, %
1	77.0	74.7	3.1
2	72.9	78.5	7.2
3	74.0	79.2	6.5
4	74.2	72.0	3.0
5	85.7	80.7	6.2

Source Roldán et al. (2016)

experimental value obtained for each test with the simulation one, the average thermal efficiency has been evaluated from the numerical thermal efficiency of each cup (Table 4.14). The comparison showed a maximum deviation of 7.2 %, a minimum one of 3 %, and an average value of 5.2 %.

Previous studies, which have also simulated high-temperature solar processes, have considered acceptable deviations of around 9 % (Ozalp and Jayakrishna 2010). Using this criterion, the results which come from the comparison between experimental and numerical data (Tables 4.13 and 4.14) are in appreciable agreement, except for the temperature measurement of absorber cup 2. The temperature variation between the cup 2 and the others is around hundreds of degrees. This result is not consistent, because the thermal distribution on the cup surface is similar to the distribution presented by the other cups. Thus, this vast difference is due to a wrong location of the thermocouple or due to a malfunction of the measuring instrument. Hence, measurement issues of the facility can be detected by using CFD simulation.

In order to counteract measurement uncertainties, the experimental thermal efficiency was calculated taking into account the fluid temperature at the receiver outlet. If the numerical thermal efficiency of the cup 2 is compared with the one experimentally evaluated, the deviation obtained is lower than 9 % (7.15 %). This fact shows that the measurement uncertainties do not have a significant influence on the experimental efficiencies.

The study of the wind influence is based on the boundary conditions of test 3 (Tables 4.8 and 4.10), whose velocity distribution is shown in Fig. 4.10, but taking into account the wind conditions (magnitude and direction) of Table 4.9. The velocity of the return air was fixed at 1.34 m/s and its temperature at 300 K. The air temperature at the absorber cup outlet and the heat transferred in the solution domain are summarised in Table 4.15.

These results show that the greater the wind magnitude, the lower the air temperature at the cup outlet because the wind enhances the heat transfer by convection between the receiver surface and the ambient air. The wind incidence is perpendicular to the receiver surface at 0°. In this case, the air outlet temperature is higher than the one reached considering other wind directions. Thus, the greater the angles of the wind direction with regard to the perpendicular direction, the lower the air temperature at the outlet. The temperature difference between the cups, which appears without the wind influence, is reduced because of the cooling effect of the wind. This is appreciable for each wind magnitude (Table 4.15).

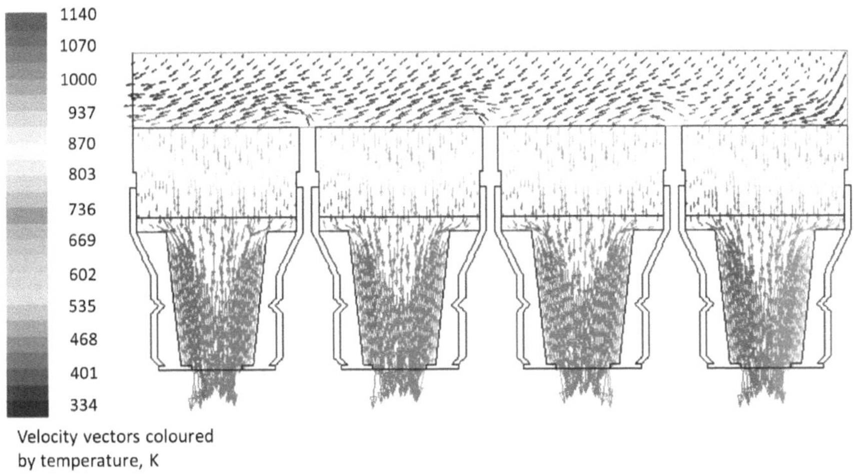

Fig. 4.10 Velocity distribution coloured by temperature in test 3 (Roldán et al. 2016)

Table 4.15 Numerical results for the analysis of the wind influence

Wind direction	v, m/s	T_{1out}, K	T_{2out}, K	T_{3out}, K	T_{4out}, K	$(T_{out})_{av}$, K
0°	0	1262	1236	1238	1261	1249
20°	3.5	1013	1013	1024	1021	1018
	7	922	921	931	916	922
45°	1.5	1029	1021	1034	1025	1027
	3.5	919	918	918	912	917
	7	778	778	789	778	781
	8	748	750	761	748	752
70°	3.5	862	863	862	858	861
	7	714	717	722	712	716
90°	1.5	996	993	990	974	988
	3.5	852	851	848	843	849
	7	702	702	704	701	702
	8	671	674	677	674	674

Source Roldán et al. (2016)

For the study of the return-air temperature, the data of test 3 have also been considered as reference conditions, the return-air velocity is set to 1.34 m/s and the three wind velocities of Table 4.12 with a direction of 45° were considered, together with the four temperatures of the return air. Figure 4.11 shows that the greater the return-air temperature (T_{ret}), the higher the air temperature at the outlet. Thus, the selection of the optimum return-air temperature is based on the heat gained by the fluid with regard to the one gained at the lower return-air temperature studied, considering only the receiver module of the facility.

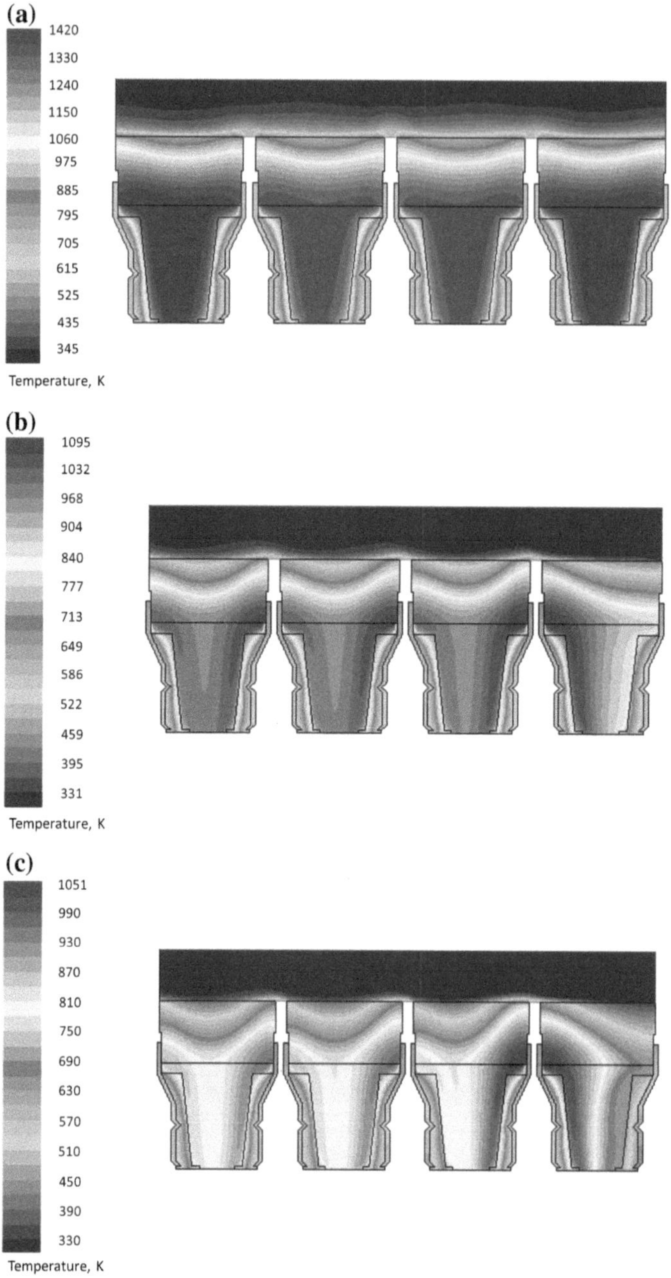

Fig. 4.11 Influence of the return-air temperature on the air temperature at the outlet with v = 0 m/s (**a**), v = 3.5 m/s and direction = 45° (**b**), v = 7 m/s and direction = 45° (**c**), including the air thermal profile for each wind condition at Tret = 600 K (Roldán et al. 2016)

For the three wind velocities selected, the optimum return-air temperature in the receiver is 400 K with a velocity of 1.34 m/s and a heat gained from T_{ret} = 300 K to T_{ret} = 400 K greater than that obtained from the others two temperature steps (receiver temperature range from T_{rec} = 400 K to T_{rec} = 500 K, and from T_{rec} = 500 K to T_{rec} = 600 K). Moreover, a higher velocity of the return air avoids hot spots in the absorber material, extending its operating time.

Figure 4.11a shows a symmetrical thermal profile because the wind velocity is zero. In this case, the thermal gradient between the central cups and the side ones

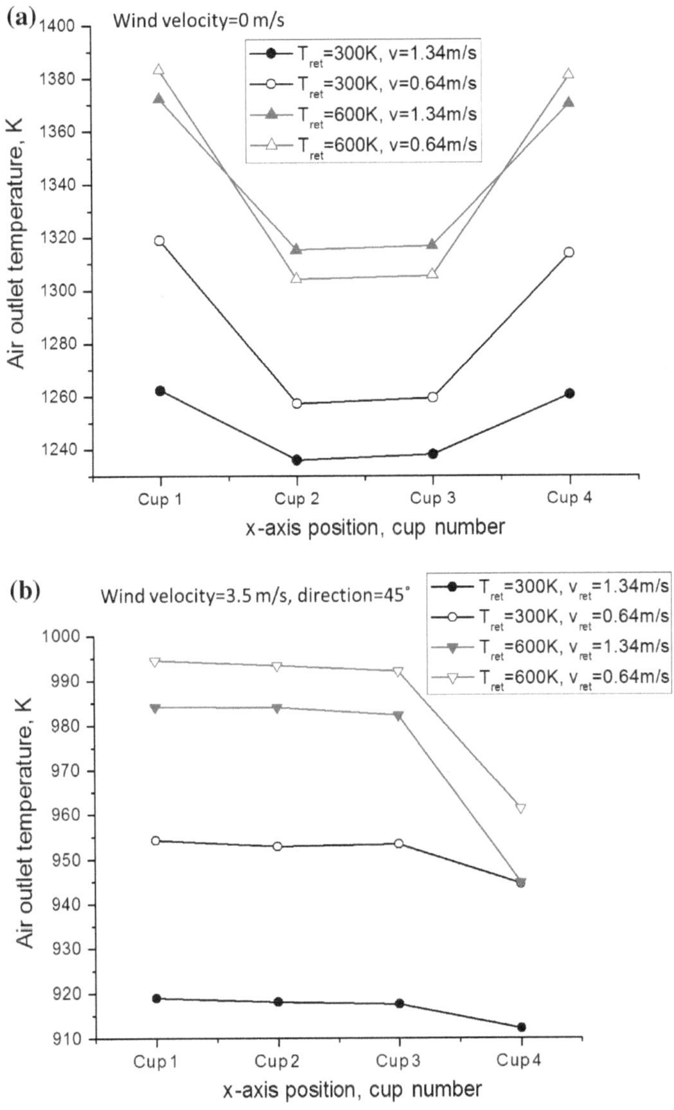

Fig. 4.12 Influence of the return-air temperature on the air temperature at the outlet with: **a** v = 0 m/s, **b** v = 3.5 m/s, direction = 45° (Roldán et al. 2016)

also increases at greater return-air temperatures. Thus, the return-air influence is greater when the wind velocity decreases and the lower the wind velocity, the greater the air temperature reached at the cup outlet. Nevertheless, the air temperature at the outlet is more homogenised at moderate wind velocities (Fig. 4.11b). Cup 4 receives the inlet wind, which explains the temperature difference between Cup 3 and Cup 4 in Fig. 4.11b, c.

Furthermore, the cooling effect is greater when the wind velocity is 7 m/s, but the return air between Cup 3 and Cup 4 is directed by the wind towards the interior of these cups. For the remaining outlets of the return air, the path flow has been homogenised in the area between Cup 1 and Cup 2. As a consequence, the outlet temperature in Cup 3 is greater than that in Cup 2. Nevertheless, when the wind velocity is 3.5 m/s, the cooling has a greater influence on Cup 4 than on the other cups and the effect of the return air on Cup 1 and Cup 2 is higher than in the case with a wind velocity of 7 m/s.

The return-air velocity was studied comparing two cases: the one with a wind velocity of 3.5 m/s and 45° of incidence, and the case without considering wind. Two different return-air velocities were selected (1.34 and 0.64 m/s) together with two values for the return-air temperatures (300 and 600 K) and the boundary conditions considered in test 3.

Figure 4.12 summarises the simulation results obtained for the cases selected. Figure 4.12a shows the data corresponding to the case without wind, where lower return-air velocities tend to reach higher air temperatures at the outlet. Nevertheless, the effect of the return air is lower for the central cups. In the other cases, the air temperature at the outlet increases with lower return-air velocities and a greater homogeneity of the air temperature in cups 3 and 4 is obtained. In Fig. 4.12b, the temperature distribution also tends to become more homogenised at moderate wind velocities.

4.5.7 Summary and Conclusions

A CFD two-dimensional model has been developed in order to study the influence of the wind and return-air conditions on the efficiency of an open volumetric receiver. This model has been validated comparing experimental and numerical data. The average deviations obtained from the evaluation of the air temperature were lower than 5 %, except for the cup 2, which presented an average deviation of around 26.4 %. After the data evaluation and considering both the model assumptions and the measuring uncertainties, it was concluded that the cup-2 deviation was due to an error in the measurement procedure, either because of a wrong thermocouple location or a thermocouple malfunction. Thus, CFD simulation has revealed a measurement issue in the experimental setup.

Furthermore, the thermal efficiency has been calculated for numerical and experimental results, obtaining a maximum deviation of 7.2 % and an average deviation of 5.2 % for an experimental thermal-efficiency range of 72–86 %. Consequently, the 2D model has been considered reliable.

The comparison between the results of a selected test and those obtained from the same case without the wind effect showed that higher temperatures at the cup outlet can be reached when there is no wind. In order to determine the influence of the wind conditions, different velocity magnitudes and incidence angles have been studied. In cases with a greater incidence angle with regard to the perpendicular direction (0°), the air temperature at the outlet decreases with increasing wind velocity. Furthermore, at a fixed velocity, the increase of the incidence angle cools the receiver and the air temperature at the absorber-cup outlet also decreases.

The evaluation of the return-air influence has regarded its temperature and velocity as parameters. If there is no wind, the air temperature at the outlet increases with increasing return-air temperature. However, this increase is not proportional to the air temperature increase between consecutive air temperatures. Thus, the thermal increase is greater between 300 and 400 K than the one obtained between the other two ranges (400 and 500 K or 500 and 600 K). This effect is also observed when there is wind.

Moreover, higher air temperatures at the absorber outlet have been reached at a lower return-air velocity (0.64 m/s). This behaviour has obtained both when there is wind and when it does not exist. In general, the influence of the return-air conditions are intensified when there is no wind. Therefore, the wind conditions are a limiting factor to enhance the thermal efficiency of the receiver. Thus, new receiver designs should avoid the direct incidence of the wind and should also consider an optimisation of the return-air conditions.

Finally, this study shows the ability of CFD simulation for supplying enough information about the system analysed with a simplified model in order to optimise its operating conditions.

References

Becker M, Fend T, Hoffschmidt B, Pitz-Paal R, Reutter O et al (2006) Theoretical and numerical investigation of flow stability in porous materials applied as volumetric solar receivers. Sol Energ 80:1241–1248. doi:10.1016/j.solener.2005.11.006

Dantzing JA, Tucker CL III (2001) Modeling in material processing. Cambridge University Press, Cambridge

Falcone PK (1986) A handbook for solar central receiver design. SAND89-8009. Sandia National Laboratories, Livermore

Haeger M, Keller L, Monterreal R, Valverde A (1994) Phoebus Technology program Solar Air receiver (TSA): operational experiences with the experimental set-up of a 2.5 MWth volumetric air receiver (TSA) at the Plataforma Solar de Almería. Project report (Ref. PSA-TR02/94)

Ho CK, Iverson BD (2014) Review of high-temperature central receiver designs for concentrating solar power. Renew Sust Energ Rev 29:835–846. doi:10.1016/j.rser.2013.08.099

Hoffschmidt B, Tellez FM, Valverde A, Fernandez J, Fernandez V (2003) Performance evaluation of the 200-kWth HiTRec-II open volumetric air receiver. J Sol Energy-T ASME 125:87–94. doi:10.1115/1.1530627

International Energy Agency (2014) Technology roadmap: solar thermal electricity. https://www.iea.org/publications/freepublications/publication/technologyroadmapsolarthermalelectricity_2014edition.pdf. Accessed 20 May 2016

Marcos MJ, Romero M, Palero S (2004) Analysis of air return alternatives for CRS type open volumetric receiver. Energy 29:677–686

Ozalp N, Jayakrishna D (2010) CFD analysis on the influence of helical carving in a vortex flow solar reactor. Int J Hydrogen Energy 35:6248–6260

Roldán MI (2013) Design and thermal analysis of a volumetric receiver system installed in a high-temperature solar furnace. ISBN 978-84-7834-696-7, CIEMAT, Madrid

Roldán MI, Fernández-Reche J (2016) CFD analysis of supercritical CO2 used as HTF in a solar tower receiver. AIP Conf Proc 1734:030031. doi:10.1063/1.4949083

Roldán MI, Smirnova O, Fend T, Casas JL, Zarza E (2014a) Thermal analysis and design of a volumetric solar absorber depending on the porosity. Renew Energ 62:116–128. doi:10.1016/j.renene.2013.06.043

Roldán MI, Valenzuela L, Fernández J (2014b) Computational fluid dynamics in cocentrating solar technologies. In: Al-Baghdadi MARS (ed) Computational fluid dynamics applica-tions in green design, 1st edn. International Energy and Environment Foundation, Iraq

Roldán MI, Fernández J, Valenzuela L, Vidal A, Zarza E (2015) CFD Modelling in solar thermal engineering. In: Al-Baghdadi MARS (ed) Engineering applications of computational fluid dynamics. International Energy and Environment Foundation, Iraq, vol 3, 1st edn

Roldán MI, Fernández J, Ballestrín J (2016) Computational fluid dynamics evaluation of the operating conditions for a volumetric receiver installed in a solar tower. Energy 94:844–856. doi:10.1016/j.energy.2015.11.035

Romero M, Téllez FM, Valverde A (1999) Operation, test and re-evaluation of PHOEBUS-TSA air volumetric receiver. Project report (Ref. TSA_99-T01-IN-C01), PSA

Chapter 5
Heat Transfer Fluids Used in Concentrating Solar Thermal Technologies

Heat transfer fluids are critical components for storing and transferring thermal energy in concentrating solar thermal technologies. Therefore, they play a key role in the overall performance and efficiency of the solar thermal electricity (STE) plants. Since a large amount of heat transfer fluid is required to operate a STE plant, it is necessary to minimise its cost while maximising its performance. In this context, this chapter aims to analyse by CFD simulation the influence of using different heat transfer fluids in solar receivers in order to improve the overall efficiency of the STE plant.

5.1 Introduction

The reduction of the levelised cost of energy leads to the achievement of a more competitive technology. In order to minimise the cost of solar thermal electricity, higher efficiency power cycles are required. Hence, alternative concepts for power cycles have been proposed such as air-Brayton, supercritical CO_2 (s–CO_2) Brayton, and ultra-supercritical steam cycles (Ho and Iverson 2014). The resulting power blocks must work at higher temperatures than those previously reached in STE plants.

Central receiver systems are based on a concentrating solar thermal (CST) technology able to achieve high temperatures. Current central receiver concepts employ either water/steam or molten nitrate salts as heat transfer fluids (HTF) in subcritical Rankine power cycles. The conversion efficiency obtained in commercial operating solar-tower plants is typically between 30 and 40 % at inlet temperatures lower than 600 °C (Ho and Iverson 2014). The key point is to reach higher temperatures in order to increase the thermal-to-electricity efficiency of the power cycle, according to the Carnot's theorem. Nevertheless, commercial fluids such as molten salts become chemically unstable at temperatures greater than 600 °C, producing oxide ions that increase the material corrosion (Bradshaw and

M.I. Roldán Serrano, *Concentrating Solar Thermal Technologies*,
Green Energy and Technology, DOI 10.1007/978-3-319-45883-0_5

Carling 1987), which results in significant mass loss and reduction of the operating lifetime of the facility.

On the other hand, parabolic trough with thermal oil as HTF is the most mature CST technology, but its working temperature is limited to 400 °C. Therefore, other fluids are being investigated such as water/steam in a direct steam generation concept. This option saves heat exchangers and some specific equipment required to ensure the quality stabilisation of the HTF, but high operating pressure forces the use of more expensive components (IEA 2014).

Since HTF is one of the most important components for overall performance of CST system, it is important to select the most appropriate working fluid according to the facility used. This implies that the selected HTF reaches high temperatures and provides an efficient and easy operation of the STE plant.

5.2 Heat Transfer Fluids for CST Systems

In previous chapters, it has been described the operating principle of CST systems. It is based on the concentration of solar radiation by mirrors that focus the beams onto a receiver, where the heat is collected by a thermal energy carrier, the so-called heat transfer fluid (HTF). This fluid can be used to directly drive a turbine to produce power or, more commonly, be combined with a heat exchanger and a secondary cycle to generate steam (Zhang et al. 2013).

Desired characteristics of a HTF include:

- Low melting point
- High boiling point and thermal stability
- Low vapour pressure (lower than 1 atm) at high temperature
- Low corrosion with metal alloys used to contain the HTF
- Low viscosity
- High thermal conductivity
- High heat capacity for energy storage
- Low cost.

The HTFs can be divided into six main groups, according to the type of material: air and other gases, water/steam, thermal oils, organics, molten salts and liquid metals. The working temperature range for organics and thermal oils are (12–393 °C) and (−20–400 °C), respectively. Molten salts is the most widely studied HTF due to their high working temperature (greater than 500 °C) and heat capacity, low vapour pressure and corrosive property, and good thermal and physical properties at high temperatures. Furthermore, liquid metals also appear to be promising candidates for high-temperature solar plants (Vignarooban et al. 2015).

As mentioned, high operating temperatures enable the improvement of the overall efficiency in the CST plant, and molten salts are the most promising HTF candidates at temperatures up to 800 °C, but the corrosion issues are increased in

CST plants operated with molten salts compared to other HTFs, mainly due to the high operating temperatures.

In order to study in more detail the different working fluids that can be used in CST technologies, the following sections describe separately the commercial HTFs and those proposed as alternative fluids to enhance the efficiency of the STE plant.

5.3 Heat Transfer Fluids Used in Commercial CST Plants

The most commonly used HTFs in commercial CST plants are thermal oils, organic materials, molten salts and water/steam. Further information about each fluid is included below.

5.3.1 Thermal Oils

Thermal oils act as thermal energy carrier mainly in parabolic trough plants. There are three types: mineral oil, silicone oil, and synthetic oil, which have almost the same thermal conductivity (0.1 W/m-K) and can be thermally stable only up to 400 °C. This is the reason because they are not used for high temperature and highly efficient solar thermal systems, and another limitation is their high cost.

Experimental performances of medium-temperature solar thermal collectors found that a type of silicon oil resulted to have better heat collection characteristics and low saturation rate in comparison to pressurised water. The main advantages of the silicon oil tested are: constant thermal conductivity over a wide range of temperature, low corrosivity and low flammability. However, increased pumping rate is required due to its low specific heat capacity which makes the process costly (Nkwetta et al. 2013).

Some thermal oils can be thermally stable up to around 300–400 °C, and Syltherm 800 is one of those that present the best thermal capacity. Furthermore, some paraffinic oils have an operating temperature range of 70–550 °C, and some aromatic oils have an operating range of 40–600 °C (Vignarooban et al. 2015).

5.3.2 Organics

The most common organic material used as HTF in CST systems is Biphenyl/Diphenyl oxide which is a eutectic mixture of two very stable organic compounds: Biphenyl ($C_{12}H_{10}$) and Diphenyl oxide ($C_{12}H_{10}O$). There are eight STE plants operating with this HTF in the world, all located in Spain. Biphenyl/Diphenyl oxide has a very narrow operating temperature range, within 12–393 °C. Furthermore, the density of this organic mixture increases with the concentration of the diphenyl ether and it has an average thermal conductivity of 0.13 W/m-K at the operating temperature range and at ambient pressure (Cabaleiro et al. 2012).

5.3.3 *Molten Salts*

As pointed out above, molten salts are good HTFs mainly due to their thermal stability at high temperatures (greater than 500 °C). They also have properties comparable to water at high temperature, such as viscosity and low vapour pressure. However, molten salts are not thermally stable at temperatures higher than 600 °C. From this temperature, this fluid acts as the electrolyte in a corrosive system in which the metal is attacked and corrosion issues appear.

On the other hand, an important advantage of using molten salts in solar towers is their capability for thermal energy storage. The first system which used molten salts as both HTF and energy storage medium was a central receiver system (Solar Two) with energy storage capability for 3 hours. The second solar tower system with this concept was Gemasolar with a production capability of 17 MW_e and 15 hours of thermal energy storage. This was the first commercial central receiver system with molten salts acting as HTF and storage medium (Vignarooban et al. 2015).

Most of the molten salts used are based in nitrates/nitrites. However, worldwide nitrate salt production is restricted. Therefore, alternative HTFs made from inexpensive and abundant materials are being investigated.

5.3.4 *Water/Steam*

Seven commercial CST plants operate with water/steam as a single fluid (mostly central receiver systems), but the main problem is the scarcity of water in desert regions where these plants are mostly located due to the large land area available and the high direct solar radiation existing there. Moreover, corrosion problems at high temperatures also appear when it is utilised water/steam in the CST facility (Vignarooban et al. 2015).

The use of water/steam as both HTF and working fluid simplifies the system because it is not necessary the heat exchange between the fluid heated by solar radiation and the one that carries the energy to the turbine. This configuration leads to improved efficiency, and also decreases the cost of electricity production.

Nevertheless, research and development of water/steam still continues because direct steam generation implies high operating pressures and complex controls for plant operation that should be better studied.

5.4 Alternative Heat Transfer Fluids in CST Plants

The improvement of the overall efficiency in a CST plant leads to the achievement of higher temperatures in the HTF, in order to integrate more efficient power cycles in the facility. In this context, it is necessary to propose alternative HTFs which are able to reach high temperatures without altering their structure, abundant, environmental friendly and with good thermal performance. Some of the alternative fluids that are being studied are: air and other gases, liquid metals, solid particles acting as HTFs and supercritical fluids.

5.4.1 Air and Other Gases

Air is not a common HTF in large STE plants. There is only a pre-commercial plant (Jülich solar tower) with a production of 1.5 MW_e which utilises air as HTF. In this case, it is possible the operation in an extensive temperature range. Thus, air at atmospheric pressure is heated up to about 700 °C in Jülich solar tower and then the hot is used to produce steam (Tian and Zhao 2013). The greatest advantage of this technology is its cost-effectiveness and high efficiencies, because of the abundance and cost-free nature of the atmospheric air.

The very low dynamic viscosity of the air compared to the one of molten salts or liquid metals gives good flow properties inside the pipelines that favour the heat transfer and may compensate its low thermal conductivity. In addition, corrosion issues have also been found in carbon steels due to their contact with air at high temperatures.

Other gases such as helium have also been investigated for use in CST plants. Helium was widely utilised as working substance in high temperature nuclear reactors and it is also relatively affordable, since it can be generated in natural gas extraction processes. Similar to air, helium can also be used at very high temperatures, but the major drawbacks are the low heat capacity and heat transfer between the fluid and the internal surfaces of the pipelines. Consequently, high pressures and high fluid velocities are required (Massidda and Varone 2007).

5.4.2 Liquid Metals

Liquid metals have been used in nuclear industries since 1940s and are being studied for their use in CST systems as HTF and thermal energy storage medium. They have not been utilised in commercial CST applications, but they have promising properties, such as extensive operating temperature range, low viscosity and efficient heat transfer characteristics. As reference, liquid sodium has an operating temperature range between 98 °C and 883 °C (Vignarooban et al. 2015).

These materials are relatively more costly than molten salts or water/steam and their heat capacity is lower than the one of nitrate/nitrite based salts, which makes them less favourable to be used as thermal storage medium.

5.4.3 Solid Particles

Solid particles can be used as HTF due to their inherent storage capacity. For example, sand-like ceramic particles are able to achieve temperatures of around 1000 °C. They can be stored in an insulated tank and/or used to heat a secondary working fluid for the power cycle.

Desirable properties of the solid particles include (Ho and Iverson 2014):

- High packing density
- High heat capacity
- Resistance to mechanical and thermal shock
- Resistance to sintering and agglomeration
- Corrosion resistance in air and other media
- High solar absorptance and low thermal emittance
- Low cost and wide availability.

Computational fluid dynamics models of falling particle receivers have been developed in order to predict the performance of this alternative receiver concept (Tan and Chen 2010).

Furthermore, innovative HTFs may include nano-fluids. Dispersing solid particles in fluids enhances thermal conductivity, but particles settle rapidly in fluids. Nano-particles, possibly improved with surfactants or stabilisers, could remain in suspension and expose a surface area per unit volume a million times larger than that of micro particles, offering improved heat transfer properties.

5.4.4 Supercritical Fluids

Supercritical fluids such as s–CO_2 can be operated at very high temperatures and are able to act as both HTF for solar collector and working fluid for the power block. In fact, s–CO_2 can achieve higher conversion efficiencies at lower temperatures in comparison with steam Rankine and helium cycles.

However, due to the high operating pressures required to maintain the supercritical condition, s–CO_2 is not appropriate for parabolic trough plants because there are extensive pipelines used. Thus, the most compatible CST technology is the central receiver system, considering fluid temperatures above 750 °C (Vignarooban et al. 2015).

In order to evaluate the thermal behaviour of s–CO_2 in a tubular receiver for a solar tower and to determine the initial operating conditions for the receiver design selected, a CFD model has been developed. The following section describes in detail the methodology used to compare the s–CO_2 behaviour with the molten-salt one, considering the same receiver design.

5.5 CFD Analysis of a Tubular Receiver Design for Two Different HTFs

This analysis has been developed because emerging receiver designs consider the achievement of higher efficiencies using advanced power cycles, such as s–CO_2 closed-loop Brayton cycles. For that purpose, it is included the direct heating of s–CO_2 in tubular receiver designs capable of withstanding high internal fluid pressures (around 20 MPa) and temperatures (900 K) (Ho and Iverson 2014). Due to the high pressures required and the presence of moving components installed in the pipelines (ball-joints and/or flexible connections), the use of s–CO_2 presents many technical challenges due to the compatibility of seal materials and fluid leakages of the moving connections. These problems can be better controlled in solar tower systems because the receiver is fixed (Vignarooban et al. 2015).

In this regard, a preliminary analysis has been developed using the design of a molten-salt tubular receiver which was previously tested. In order to minimise the computational cost, a simplified geometry has been considered in the model. CFD simulations have been carried out to determine the minimum operating conditions required for the use of s–CO_2 as HTF and to compare the efficiencies reached by both fluids.

5.5.1 Receiver Description

The selected receiver consisted of three different modules through which the HTF flows consecutively. Each module is made of 20 alloy tubes that receive the solar radiation coming from the heliostat field and transfer the heat to the working fluid. Figure 5.1a shows the fluid flowing through the central module and then it passes through the lateral ones. There is a manifold at the inlet and outlet of each module to homogenize the flow conditions.

5.5.2 Procedure

In order to develop a preliminary study of the s–CO_2 behaviour in comparison with the molten-salt one, a 2D symmetrical simulation domain of the module 2 (Fig. 5.1b) has been considered. This initial approach was regarded to minimise the computational cost and to obtain initial results which lead the study towards the most interesting aspects that will be analysed in future investigations.

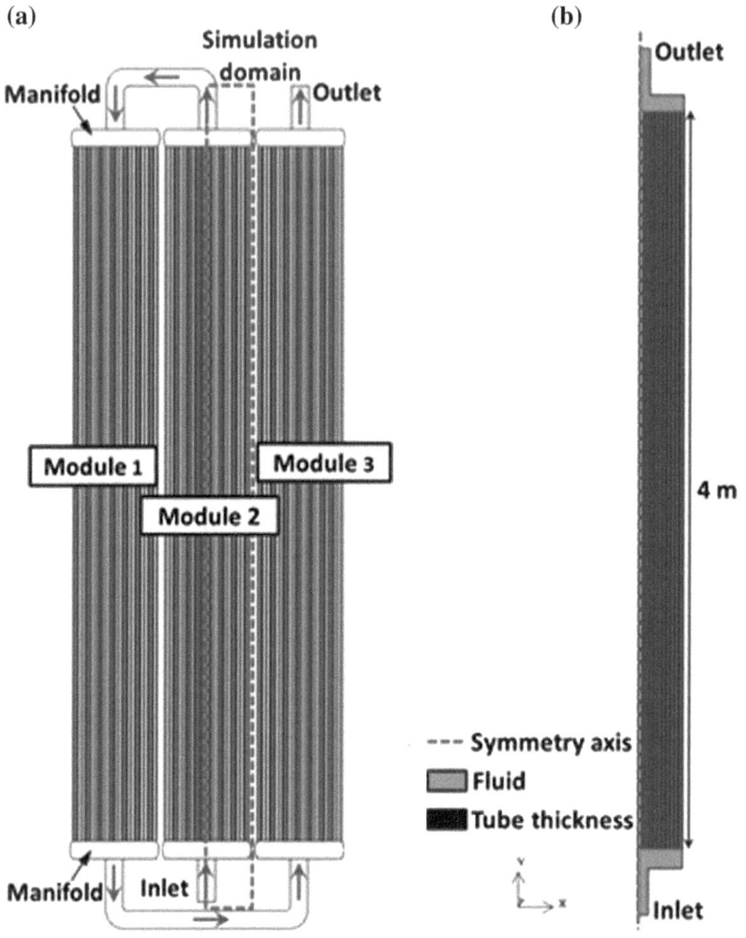

Fig. 5.1 Geometry: **a** scheme of the tubular receiver, **b** simulation domain (Roldán and Fernández-Reche 2016)

The CFD model created was validated by experimental data obtained from tests of the molten salt receiver, taking into account the measurements at a quasi-steady state selected. The measured irradiance distribution was considered as boundary condition and the pressure and mass flow at the inlet are those measured at the pump impulsion. The fluid temperature is measured at the inlet and outlet of each module and the pressure is used as boundary condition.

Furthermore, simulations with s–CO_2 as HTF were developed at several different fluid inlet temperatures (800, 715, 600 and 500 K) to study the operating conditions required for each case. Table 5.1 summarises the steady-state conditions for each fluid.

Table 5.1 Steady-state conditions

Fluid	Inlet pressure, Pa	Outlet pressure, Pa	Fluid temperature at the inlet, K	Total power received, kW
Molten salts	589000	347000	715	720
s–CO_2	11000000	7500000	800, 715, 600, 500	720

Source Roldán and Fernández-Reche (2016)

5.5.3 CFD Modelling

The solution domain selected was a two-dimensional section of the receiver central module with the following zones: inlet pipe, inlet manifold, receiver tubes, outlet manifold and outlet pipe. Regarding a symmetry condition for this geometry, the domain consisted of two sub-domains: the thickness of each receiver tube and the fluid (Fig. 5.1b).

As mentioned in Chap. 3, conservation laws determine the fluid dynamics behaviour; therefore, the proposed CFD model has required solving the continuity, momentum and energy equations adapted to the steady-state flow condition. Furthermore, the selected mesh is built of quadrilateral cells, whose equiangle-skew value is included in the range of 0–0.44. It means that the mesh quality is appropriate because all grid elements are within the good quality range.

The material properties considered were obtained from literature and the boundary conditions selected were: inlet/outlet pressure, natural convection in outer walls because the insulation does not eliminate heat losses to the receiver cavity, symmetry for the cross-cutting axis and the contact walls are coupled with the linked areas (Fig. 5.2b).

Furthermore, the volumetric heat source for each absorber tube was obtained from the analysis of the heat-flux distribution measured, showing an average irradiance of 122 kW/m^2. In order to define the tube zone with the maximum heat flux, each tube length of the receiver was divided into three parts according to the heat-flux distribution. The volumetric heat source was implemented in each zone depending on its volume (404651, 1540465 and 845581 W/m^3 for zones A, B and C, respectively).

Both the definition of a 2D symmetrical geometry and the approach of a constant heat source for each tube zone reduce the computational cost. Moreover, the boundary conditions were adapted to the HTF used. The outlet pressure was set at 0.35 MPa for molten salts (measured value) and 7.50 MPa for s–CO_2 (critical pressure of 7.38 MPa).

The viscosity model has been selected according to the Reynolds number (Re) evaluated for molten salts in three zones: receiver tube, manifold and module inlet/outlet. The Re value was between 6.12×10^5 and 9×10^6 which belongs to a turbulent regime. Nevertheless, κ-ε renormalization group (RNG) model was used to take into account areas with low Re.

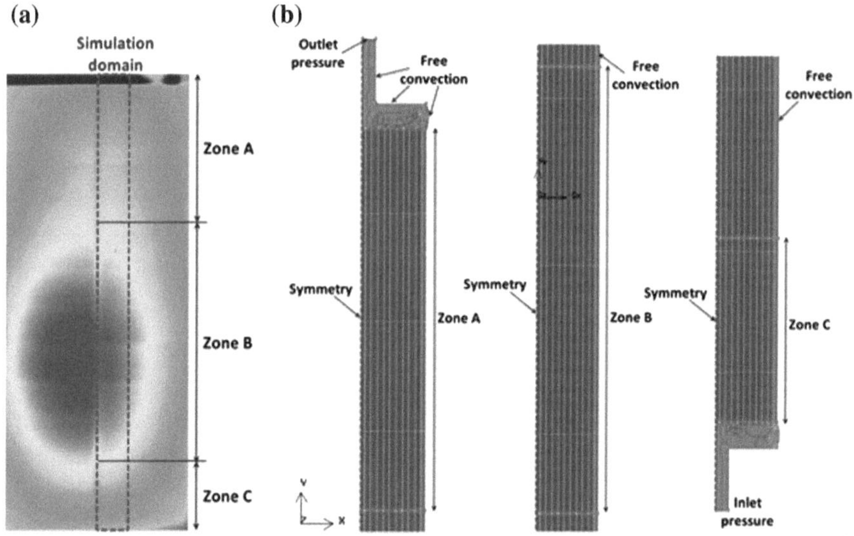

Fig. 5.2 Definition of the boundary conditions: **a** heat flux distribution, **b** boundary conditions in the solution domain (Roldán and Fernández-Reche 2016)

5.5.4 *Results and Discussion*

The CFD model was validated taking into account a molten-salt test. The variable considered was the absolute inlet pressure of the fluid at the steady state selected. In this case, the outlet pressure was fixed as boundary condition. The deviation between experimental and numerical data is summarised in Table 5.2. According to previous studies about simulation of high-temperature solar processes which accepted deviations of around 9 % (Ozalp and Jayakrishna 2010), numerical results are in agreement with experimental data, since the deviation is lower than 7 %.

In order to evaluate the s–CO_2 behaviour as HTF and determine its initial operating conditions, simulations with different s–CO_2 inlet temperatures were developed. The outlet pressure was set at 7.5 MPa in a first approach because further studies could take into account the three receiver modules to define this parameter. Thus, the required inlet pressure was around 9.67 ± 0.05 Pa, and the temperature reached by the fluid is included in Table 5.3.

The largest increase in heat gained by the fluid was obtained at the inlet temperature of 800 K. However, further research has been proposed to evaluate the radiation heat losses in addition to the convection losses considered for this case. Thus, simulation data for 715 K were considered in a more conservative approach to analyse the initial operating conditions required for s–CO_2. Figure 5.3 shows a non-homogeneous distribution both for the fluid temperature (Fig. 5.3a) and for the

Table 5.2 Comparison between experimental and numerical data for model validation

Experimental inlet pressure, Pa	Numerical inlet pressure, Pa	Deviation, %
589000	550023	6.62

Source Roldán and Fernández-Reche (2016)

Table 5.3 Simulation results for s–CO_2

Simulation	Inlet temperature, K	Outlet temperature, K	Heat gained by the fluid, kW	Maximum temperature reached, K
1	800	812	358.27	1038
2	715	726	321.18	903
3	600	608	226.49	758
4	500	506	167.46	607

Source Roldán and Fernández-Reche (2016)

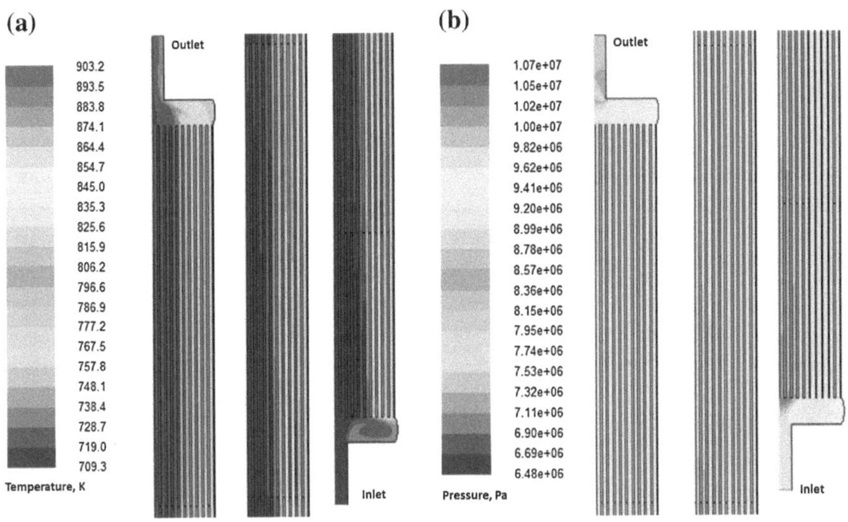

Fig. 5.3 Simulation results: **a** Thermal distribution of s–CO_2 (T_{inlet} = 715 K), **b** Pressure distribution of s–CO_2 (T_{inlet} = 715 K) (Roldán and Fernández-Reche 2016)

pressure (Fig. 5.3b). Because of this fact, the receiver design requires to be adapted to the supercritical fluid in order to obtain more homogeneous operating conditions. Hence, the analysis of several different receiver configurations should be proposed for future work.

Simulation results obtained from the use of s–CO_2 as HTF, considering the inlet temperature of 715 K, are summarised in Table 5.4.

Table 5.4 Simulation results for s–CO_2

Fluid	Inlet pressure, Pa	Outlet pressure, Pa	Mass flow, kg/s	Maximum temperature reached, K	Heat gained by the fluid, kW
s–CO_2	9695240	7500000	24.90	903	321.18

Source Roldán and Fernández-Reche (2016)

The operating pressure for s–CO_2 is much higher than the one required by the molten salts (Table 5.2). Nevertheless, the pressure required is achievable by equipment used in other solar facilities (Muñoz-Anton et al. 2014). Furthermore, the maximum temperature reached is higher than the upper working-temperature limit for molten salts (873 K). As a consequence, the use of s–CO_2 as HTF in solar tower receivers can be a promising alternative, but taking into account the high operating pressure and temperature required for its use in the solar facility.

5.5.5 *Summary and Conclusions*

A simplified CFD model has been developed in order to obtain a preliminary analysis of a tubular solar receiver considering s–CO_2 as HTF. Thus, the simulation domain selected consisted of a 2D symmetrical geometry of the central receiver module. The model validation was carried out by the comparison between experimental data coming from a tested molten-salt tubular receiver and numerical results, regarding the operating conditions at the steady-state selected.

The validation took into account measurements of fluid pressure at the inlet of the central receiver module (simulation domain). The comparison between these data showed a maximum deviation lower than 7 %. Hence, the CFD model was used to study the s–CO_2 behaviour for several different inlet temperatures.

The s–CO_2 inlet temperature of 715 K was set to determine the minimum operating conditions to maintain the supercritical phase. The operating pressure for s–CO_2 was much higher than the one required by the molten salts, but it could be achievable by equipment used in solar facilities for parabolic-trough collectors. Furthermore, the maximum temperature reached was higher than the upper working-temperature limit for molten salts (873 K). Thus, the use of s–CO_2 as HTF in solar tower receivers appears to be a promising alternative, taking into account both the operating conditions required and their maintenance cost.

Furthermore, simulation results show that the design of the tubular receiver regarded does not allow homogeneous thermal and pressure profiles. Thus, further CFD simulations should be required to optimise the receiver design.

References

Bradshaw RW, Carling RW (1987) A review of the chemical and physical-properties of molten alkali nitrate salts and their effect on materials for solar central receivers. J Electrochem Soc 134(8B):C510–C511

Cabaleiro D, Pastoriza-Gallego MJ, Pineiro MM, Legido JL, Lugo L (2012) Thermophysical properties of (diphenyl ether + biphenyl) mixtures for their use as heat transfer fluids. J Chem Thermodyn 50:80–88. doi:10.1016/j.jct.2012.02.001

Ho CK, Iverson BD (2014) Review of high-temperature central receiver designs for concentrating solar power. Renew Sustain Energy Rev 29:835–846. doi:10.1016/j.rser.2013.08.099

International Energy Agency (2014) Technology roadmap: Solar thermal electricity. https://www.iea.org/publications/freepublications/publication/technologyroadmapsolarthermalelectricity_2014edition.pdf. Accessed 20 May 2016

Massidda L, Varone A (2007) A numerical analysis of a high temperature solar collecting tube, using helium as a heat transfer fluid. CRS4 report

Muñoz-Anton J, Biencinto M, Zarza E, Díez LE (2014) Theoretical basis and experimental facility for parabolic trough collectors at high temperature using gas as heat transfer fluid. Apl Energy 135:373–381. doi:10.1016/j.apenergy.2014.08.099

Nkwetta DN, Smyth M, Zacharopoulos A, Hyde T (2013) Experimental performance analysis and optimization of medium temperature solar thermal collectors with silicon oil as a heat transfer fluid. Int J Energy Res 37:570–581

Ozalp N, JayaKrishna D (2010) CFD analysis on the influence of helical carving in a vortex flow solar reactor. Int J Hydrogen Energy 35:6248–6260. doi:10.1016/j.ijhydene.2010.03.100

Roldán MI, Fernández-Reche J (2016) CFD analysis of supercritical CO_2 used as HTF in a solar tower receiver. AIP Conf Proc 1734:030031. doi:10.1063/1.4949083

Tan TD, Chen YT (2010) Review of study on solid particle solar receivers. Renew Sustain Energy Rev 14(1):265–276. doi:10.1016/j.rser.2009.05.012

Tian Y, Zhao CY (2013) A review of solar collectors and thermal energy storage in solar thermal applications. Appl Energy 104:538–553. doi:10.1016/j.apenergy.2012.11.051

Vignarooban K, Xu X, Arvay A, Hsu K, Kannan AM (2015) Heat transfer fluids for concentrating solar power systems—a review. Appl Energy 146:383–396. doi:10.1016/j.apenergy.2015.01.125

Zhang HL, Baeyens J, Degreve J, Caceres G (2013) Concentrated solar power plants: review and design methodology. Renew Sustain Energy Rev 22:466–481. doi:10.1016/j.rser.2013.01.032

Chapter 6
Conclusions

This chapter summarises the main findings of this book. It has been analysed the current situation of the energy demand and the role that solar thermal electricity plays in the demand supply. Furthermore, the most important aspects for the improvement of concentrating solar thermal technologies (more efficient solar receivers and heat transfer fluids) have been studied by using Computational Fluid Dynamics, which is presented as a versatile and powerful simulation tool.

6.1 Summary and Conclusions

The rapid growth of the world population has led to the development of an unsustainable energy supply. Current power-generation trends increase energy-related greenhouse-gas (GHG) emissions and, as a consequence, the climate degradation with an average 6 °C global warming. Thus, in the last decades, a global effort has emerged in order to face this problem.

The overall aim is to advance global development and uptake of key technologies to limit the global mean temperature increase to 26 °C in the long term. For that purpose, a range of technology solutions have been identified to prevent the increase of GHG emissions, such as greater energy efficiency, use of renewable energy and the near-decarbonisation of fossil fuel-cased power generation.

In this context, solar thermal electricity generates power without GHG emissions, and it is presented as a key technology for preventing climate change. Longer-term climate change mitigation studies tend to show significantly higher concentrating solar thermal (CST) technology deployment beyond 2050. However, this technology still requires research aimed at improving efficiency and reducing investment and power-generation costs.

The overall efficiency of the CST facility can be enhanced by reaching higher working temperatures, which implies greater thermal losses of the receiver and costly materials that withstand heavier operating conditions. Therefore, the

M.I. Roldán Serrano, *Concentrating Solar Thermal Technologies*,
Green Energy and Technology, DOI 10.1007/978-3-319-45883-0_6

appropriate selection of the heat transfer fluid (HTF) together with the optimisation of both solar receiver design and its operating conditions are essential in order to improve the efficiency of the facility.

In this context, Computational Fluid Dynamics (CFD) is presented as a versatile and powerful simulation tool able to develop a wide variety of studies by which the thermal and dynamic behaviour of a fluid can be analysed. It has been remark the application of CFD in green engineering. This is an emerging field that is gaining in importance due to the current trends in sustainable energy production.

Hence, different CFD simulations applied to CST technology have been developed in order to optimise several solar receiver designs, their operating conditions and also to study the thermal behaviour of innovative HTFs in comparison with conventional ones. Detailed descriptions of the CFD model adapted to each case have been given in the main chapters of this book (Chaps. 4 and 5) and the most important findings have been highlighted.

In CFD modelling, it is highly recommended to take into account a validation process because of the assumptions implemented to minimise the computational requirements. However, as demonstrated, the use of CFD simulation in the design and optimisation of both CST technologies and their associated engineering provides rapid information of the physical phenomena involved in the problem analysed as well as a reduction of the investment cost in the research developed.

FSC
www.fsc.org
MIX
Papier | Fördert
gute Waldnutzung
FSC® C083411

Kolibri 360 GmbH
Ferdinand-Jühlke-Straße 7
99095 Erfurt, Deutschland
produktsicherheit@kolibri360.de